Mississippi Weather and Climate

Mississippi Weather and Climate

KATHLEEN SHERMAN-MORRIS

CHARLES L. WAX

MICHAEL E. BROWN

University Press of Mississippi ⚬ Jackson

www.upress.state.ms.us

The University Press of Mississippi is a member of the Association of American University Presses.

Photographs on page II (from left to right): damage from Hurricane Camille in Biloxi, Mississippi (photo credit: National Oceanic and Atmospheric Administration, Department of Commerce); a refugee camp at Vicksburg, Mississippi, during the flood of 1927 (photo credit: Steve Nicklas, National Oceanic and Atmospheric Administration, Department of Commerce); the former Yazoo and Mississippi Valley Railroad Depot Building in Vicksburg, which was undergoing renovation, was surrounded by floodwater in May 2011 (photo credit: K. Sherman-Morris)

First printing 2012

∞

Library of Congress Cataloging-in-Publication Data

Sherman-Morris, Kathleen.
Mississippi weather and climate / Kathleen Sherman-Morris, Charles L. Wax, Michael E. Brown.
p. cm.
Includes bibliographical references and index.
ISBN 978-1-61703-260-8 (cloth : alk. paper) — ISBN 978-1-61703-261-5 (ebook) 1. Mississippi—Climate. 2. Meteorology—Mississippi. I. Wax, Charles Larry, 1946– II. Brown, Michael E., 1967– III. Title.
QC984.M7S54 2012
551.609762—dc23   2011038801

British Library Cataloging-in-Publication Data available

CONTENTS

FIGURES

TABLES

Mississippi Weather and Climate

1. INTRODUCTION

Before we get too far into what causes this or that aspect of weather or climate, it is a good idea to understand the difference between the two. Weather is the day-to-day changes in air pressure, cloud cover, temperature, humidity, and so on. Weather is what we experience when we go outside and what we watch on the local news. Climate is the average weather for a particular place. Each day has certain characteristics that, when averaged over a certain period of time, provide a picture of that place's climate. The normal high and low temperatures for a given day are climatological averages. Climate also takes into consideration the extremes a location experiences. Climatology is the study of climate, and sometimes the study can have a particular focus. For instance, tornado climatology would explain where tornadoes have historically hit the state and when.

Ask someone about weather in Mississippi and they might mention the heat or the humidity, especially if you ask them in summer. While dominated by summer conditions, Mississippi weather experiences a good deal of yearly, seasonal, and even daily variation, including some fairly extreme weather. Hurricanes and tornadoes aside, typical weather conditions have experienced some wide swings. For instance, the temperature in the state has ranged from 115°F in Holly Springs to –19°F in Corinth. Precipitation has also varied recently from very dry to very moist. In 2007, across the state there was an average of only 10 inches of precipitation from January to May. In 2009, the same 5 months experienced 39 inches. Both of these extremes have less than a 5% chance of happening in any year. It's not unusual for Mississippi residents to have to run their air conditioners one day and their heaters the next.

There are many factors that contribute to the variation in weather and climate, but a dominant control is the mid-latitude cyclone in which weather over hundreds of miles is controlled largely by the movement of air masses and pressure systems. Researchers have estimated that 75% of the daily weather experienced in Mississippi is related to the passage of frontal systems. The cyclone includes many of the components you see on the graphics in the local weathercast or online at some of the weather websites. Figure 1.1 is an example of a forecast graphic created by the Hydrometeorological Prediction Center in

Camp Springs, Maryland. On this map, you can see several prominent weather features. There is an area of low pressure (L) over Ohio associated with two fronts. Low pressure is associated with rising air. When air rises in the Northern Hemisphere, it rotates counterclockwise. These systematic features and traits of the mid-latitude cyclone are what cause the anticipated patterns of weather in Mississippi.

Typically when a mid-latitude cyclone is approaching Mississippi from the northwest, the first indication is the appearance of cirrus clouds from the southwest. At the same time, jet contrails are apparent in the sky over the state. These occurrences mean that the warm front of the cyclone is heading from the southwest toward Mississippi and it will probably rain in a day or two. The first rain is associated with the warm front and is gentle, widespread, and persistent. Following the passage of the warm front, the state is in the warm sector of the cyclone, characterized by warm, humid conditions, cumulus clouds, and winds from the south. On these days in the winter, afternoon temperatures may reach the 70s or even 80s, and condensation will accumulate on streets and in carports. A day or so later, the cold front will come across the state, usually marked by a line of thunderstorms with rapidly falling temperatures and winds shifting to the north behind the front. Then high pressure builds over the state from the northwest and clear weather dominates for a few days before the cycle begins anew.

Figure 1.1 illustrates these features. The counterclockwise rotation around the low pressure in Ohio is creating northwest winds that are transporting cold, continental air from Canada and the Midwest toward Mississippi. The leading edge of that colder air is the cold front. That cold front was expected to reach Mississippi soon. Also stretching out to the east of the center of low pressure is the warm front, created by that same counterclockwise rotation around the low that is creating southwest winds and pulling warm, maritime air up over Mississippi. The triangles of the cold front and the half-circles of the warm front indicate the direction in which the different air masses are heading, so if you follow the circulation around the low, you can see that the warm front has already made its way through Mississippi. Warm fronts do not mark abrupt changes like cold fronts do, though. For the few days before this front passed through Mississippi, the temperature gradually increased until the passage of the cold front. Behind the cold front is high pressure (H), associated with sinking air that rotates clockwise in the Northern Hemisphere, and at the surface the air flows outward, away from the center of the high. Because the air is sinking, that high pressure brings fair weather.

Because the jet stream travels further south during winter, Mississippi experiences more low pressure systems with this classic combination of cold front and warm front. This is the reason why the temperature fluctuates so much during winter and transitional seasons. During summer, the jet stream retreats

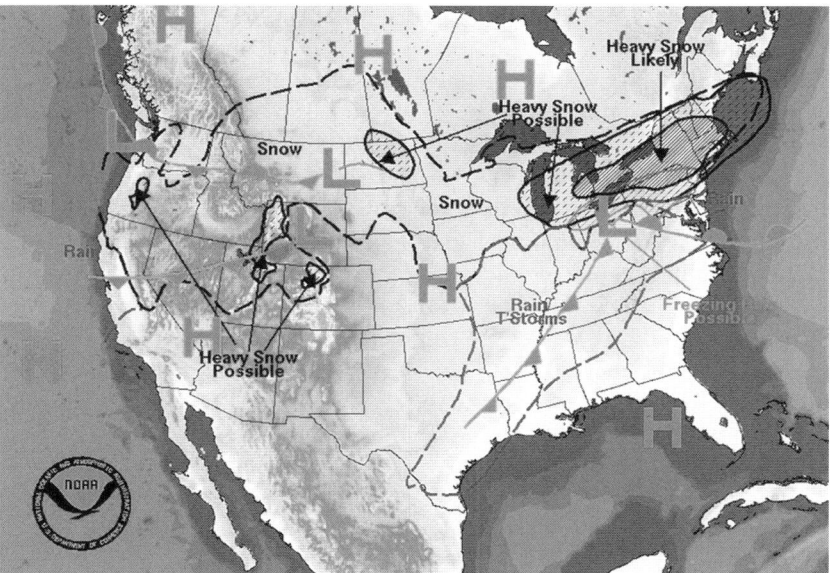

Fig. 1.1. Surface weather map. Forecast made by the Hydrometeorological Prediction Center. Image credit: National Oceanic and Atmospheric Administration, Department of Commerce

further north; as a result, we do not experience cold fronts of the same frequency or intensity. Of course, the reason the jet stream retreats further north in summer is because there is not such a pronounced change in temperature across North America. With the rest of the United States warm during summer, the relatively cooler polar air would have to travel over a large area of warm surface air, and it would therefore lose much of its cooling power by the time it arrived in Mississippi. That is why when we experience a summertime cold front, it does not really make it cold; it simply changes the direction the air is coming from. This makes the air drier, though, which can help us feel cooler in summer.

The main factors that contribute to an area's climate are latitude, elevation, and location in relation to a landmass or regional circulations. (These will be discussed in greater detail in relation to Mississippi's climate in chapter 2.) Some of these factors are more or less important than others in determining Mississippi's climate. For example, topography does not have as strong an influence on our weather and climate as it does in states with more pronounced changes in elevation, such as Washington and Oregon. The Coast Ranges and the Cascades clearly divide those states into very wet coastal regions with desert conditions inland. Mississippi's geography does play a role in its climate, however. Small changes in topography or land surface are sometimes great enough to influence weather on a smaller scale, such as where thunderstorms form.

UNDERSTANDING MISSISSIPPI'S WEATHER AND CLIMATE

William Faulkner, who was born and raised in northern Mississippi, has been quoted saying, "To understand the world, you must first understand a place like Mississippi." This book takes a closer look at the state's weather and climate. In the following chapters we expect you may learn a little bit about the state, but many of the discussions will help you to better understand what makes the weather in *this* place both unique and very much like weather in other places. In that sense, you may better understand the world by the time you are finished.

In chapter 2, we examine the reasons behind Mississippi's climate. We describe how the Gulf of Mexico, the continent of North America, and Mississippi's latitude contribute to the state's particular climate type, which is considered humid subtropical. The variation Mississippi experiences from year to year and from one season to the next is discussed, and we provide an overview of some of the weather Mississippi experiences.

The important weather elements of temperature and moisture are discussed in chapters 3 and 4. We describe seasonal and statewide variation, as well as how Mississippi compares to the rest of the country. Chapter 4 covers some of the fundamentals of atmospheric moisture, including condensation, humidity, clouds and cloud formation, and fog. We describe common settings for fog in Mississippi, as well as the seasonal and geographic patterns of precipitation throughout the state. Chapter 4 also discusses floods, which are sometimes the result of too much precipitation in Mississippi and sometimes the result of weather systems further upstream. Finally, in both chapters 3 and 4 we present and discuss temperature and precipitation averages and records.

Building on this information, chapter 5 examines thunderstorms and severe weather, including tornadoes. We also describe historic tornadoes such as the great Natchez tornado of 1840, the Dixie outbreak of 1908, the Tupelo tornado of 1936, and the Mississippi Delta tornado of 1971, as well as tornado safety.

Hurricanes have brought Mississippi misery and notoriety in recent years and are an important part of the state's weather and climate. In chapter 6, we examine the ingredients essential to their formation, their life cycle, and their structure. The seasonal forecasting of hurricanes is introduced by highlighting the methods of predicting hurricanes used by two famous predictors, Philip Klotzbach and William Gray. We then discuss the impacts of hurricanes, including storm surge, inland flooding, tornadoes, and, of course, wind. Chapter 6 describes the seasonal and geographic distribution of hurricanes. Also, significant Mississippi hurricanes are summarized, including the damage associated with them and large-scale weather conditions that favored their development. The Saffir-Simpson Hurricane Wind Scale for describing hurricane intensity will be highlighted. We also discuss hurricane safety and how to prepare for these extreme storms.

Chapter 7 focuses on winter weather, including snow and ice storms. We discuss the very slight changes in the temperature characteristics of the atmosphere that can lead to snowfall, ice, or just days of cold rain. We also present the climatology of winter weather, including who gets the most snow across the state. Historical winter weather events are also discussed, as well as safety when dealing with winter weather.

Chapter 8 describes long-term changes in climate, and chapter 9 focuses on how climate impacts aspects of our lives and the state, such as agriculture, casino tourism, and architecture. Catfish farming provides an interesting case study in the use of climate variables to conserve water and raise catfish more economically. In chapter 9, we also provide useful information for the backyard gardener, such as maps of first and last likely freeze dates and the frost-free period.

Chapter 10 focuses on delivering weather forecasts, beginning with a discussion of local television weathercasting. A few television meteorologists share their perspectives on the changing field of meteorology. This is followed by a discussion of weather folklore and myths. We describe how forecasts are made, including how weather data are collected and Mississippi weather observation networks. Finally, we discuss the educational opportunities in meteorology at Mississippi State University and Jackson State University.

In addition, each chapter has text boxes, in which we highlight one subject or discuss it further. These may describe a specific weather event, give an eye-witness account, or to relate the topic to a subject a bit outside of the realm of weather and climate.

Finally, it's important to share some information about the data used in this book. Much of the weather data are freely available from several sources. Mississippi is covered by four National Weather Service forecast offices: Memphis, Tennessee; Jackson, Mississippi; Mobile, Alabama/Pensacola, Florida; and New Orleans/Baton Rouge, Louisiana. These offices post links to some information on their websites, including highs and lows, hourly reports of weather conditions throughout the state, and some past weather conditions. Other available resources include the National Climatic Data Center and the Southern Regional Climate Center. The Hydrometeorological Prediction Center archives images of its forecast products as well as maps of fronts and pressure systems. Several private companies also make current weather information freely available. We have tried to collect the most useful information and combine it in a format that is easy to read and useful to you. We hope you enjoy the book and learn a lot about Mississippi weather and climate.

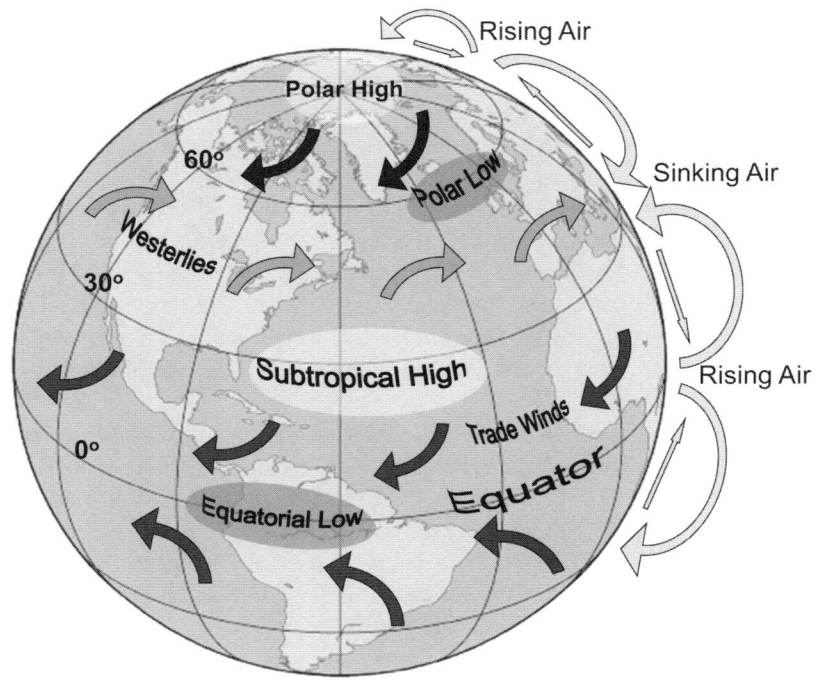

Fig. 2.1. Global circulation patterns. Blank globe from MapAbility.com

2. THE CLIMATE OF MISSISSIPPI

Numerous factors influence the climate of a region, including its latitude, its location relative to water or whether it is in the middle of a large landmass, its location on the western or eastern side of a continent, the surrounding topography, and its elevation. The exact combination of these factors gives each place its unique climate and controls the range of possible daily weather. They explain why Phoenix, Arizona, is hot and dry, whereas Buffalo, New York, sees so much snow in winter. Climatological factors account for California's Santa Ana winds and the tornado season of the Great Plains. Like these other locations, Mississippi also has its own unique set of characteristics. The climate of Mississippi is controlled primarily by three factors: its subtropical latitude, the Gulf of Mexico to the south, and the landmass of the North American continent to the north.

You can think of this chapter as your seat in Climatology 101 on the day the lecture was about Mississippi. We describe how each climatological factor plays a role in giving Mississippi its seasonal variations and average weather conditions.

LATITUDE

Mississippi's latitude ranges from 35°N at its northern border with Tennessee down to almost 30°N at the coast. This puts it in an area known as the middle latitudes. The middle latitudes are neither warm all year like the tropics, which range from 23.5° both north and south of the equator, nor cold most of the year like the polar regions, which begin at 66.5° north and south. The middle latitudes are also the area of the traveling mid-latitude cyclone, transient areas of low pressure found between 23.5 and 66.5° latitude. A cyclone is a low pressure system, which rotates counterclockwise in the Northern Hemisphere. Tropical cyclones are another type of cyclone and will be discussed in the chapter on hurricanes. Winds in the middle latitudes predominantly blow from the west, and this is the general direction from which mid-latitude cyclones approach the state.

Latitude is also important because more stationary areas of high and low pressure exist at certain latitudes, at least seasonally. These semi-permanent areas of high and low pressure form due to temperature differences across the

Earth's surface. The equator receives the most direct sunlight throughout the year, which causes the average temperature at the equator to be higher than locations farther north and south. Warm air is less dense than cold air, and when the air gets heated at the equator, it rises. The rising air forms a belt of low pressure at the equator. This air does not rise indefinitely, but begins to spread poleward once it reaches the top of the troposphere, which is the zone of the atmosphere where all weather takes place. The air that began by rising over the equator eventually begins to sink toward the Earth's surface at approximately 30° north and south latitudes. The air that subsides is generally dry and stable, which means it is not likely to rise again to form precipitation. Two semi-permanent high pressure areas that influence weather in the United States are the Hawaiian High in the Pacific Ocean and the Bermuda-Azores High in the Atlantic Ocean.

Semi-permanent high pressure areas are often associated with some of the world's largest deserts, including the Sahara and Kalahari in Africa and the Great Sandy Desert in Australia. Except for those stretches in summer when the temperature tops 100°F and cracks begin to form in the clay soil from lack of rain, Mississippi's climate is very little like that of the Sahara, even though they are both close to an area of high pressure. Why is Mississippi not as dry? Cyclones rotate counterclockwise in the Northern Hemisphere, whereas high pressure anti-cyclones rotate clockwise. So, locations on the western edge of this clockwise circulation around high pressure do not experience the same degree of subsidence (sinking air) as locations on the eastern fringes. In addition, the Gulf of Mexico provides an ample source of moisture almost year round, and the Bermuda High pumps this moisture into Mississippi, especially in summer.

The location and seasonal intensity of the Bermuda High can dominate an entire season in the state. Its influence is most pronounced during summer in Mississippi. Where the Bermuda High is located determines whether the flow of air across the state is predominantly from over the Gulf of Mexico or from over land. When placement of the high is far enough west to allow for warm, moist air to flow into the southern United States from the Gulf, but not too far west that the high sits directly over the southern United States, conditions are just right for the most hazy, hot, and muggy summer days. When the position of the Bermuda High prevents flow from over the Gulf of Mexico, such as when it is very strong and extends far to the west, the weather is typically drier. The Bermuda High, being a high pressure system, also suppresses storm development. How far westward the clockwise flow around the high extends can also influence where hurricanes track. When the flow extends farther to the west, hurricanes can be steered toward the Gulf Coast. When the Bermuda High is weaker and does not extend as far to the west, more hurricanes are directed around its circulation up the East Coast or out into the Atlantic.

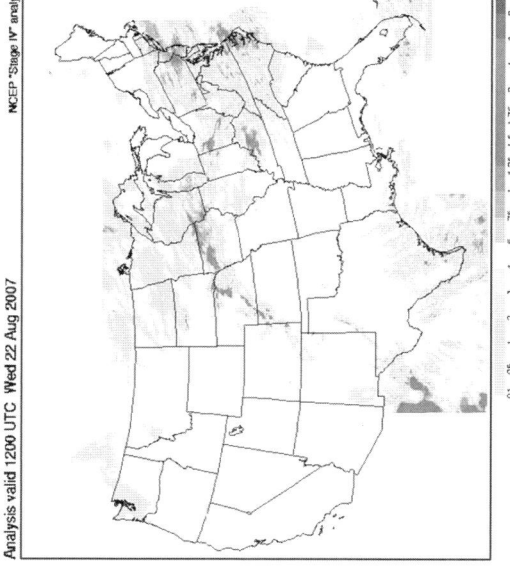
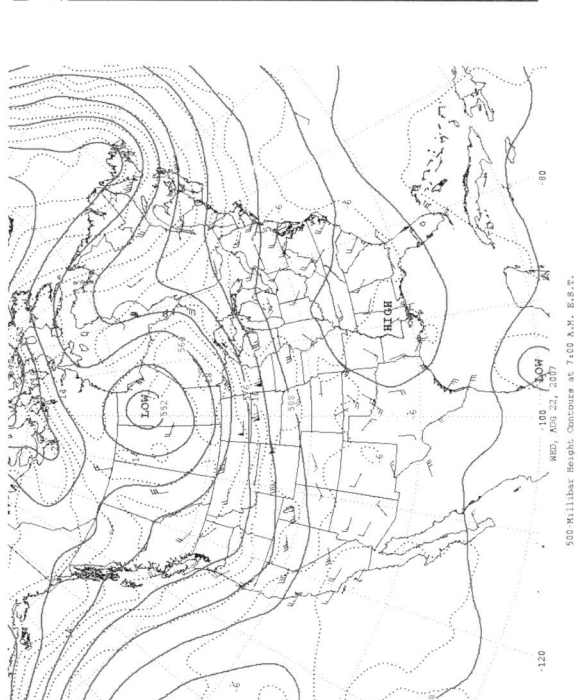

Fig. 2.2. (left) Bermuda High at 500 millibars (a unit of atmospheric pressure). A ridge of high pressure associated with the Bermuda High stretches across the Southeast and is centered over Mississippi. (right) Precipitation on that date. Notice how precipitation only formed on the fringes of the high and especially where the flow around the high was onshore (southern Texas and southern Florida).

These images were from August 22, 2007. The Bermuda High controlled the state's weather for much of summer 2007, leading to one of the driest summers on record. Images credit: National Oceanic and Atmospheric Administration /Department of Commerce available through the UCAR image archive (http://www.mmm.ucar.edu/imagearchive)

THE GULF OF MEXICO

Bodies of water can have multiple effects on a location's climate. Not only does air have circulation patterns, oceans have circulation patterns as well. Ocean currents help to equalize the imbalance in heat energy between the equator and the poles. Each ocean basin has a circulation system that transports warm water poleward along the eastern coasts of the continents surrounding it and colder water toward the equator along the western coasts of the continents. For example, California is located along the western coast of North America and therefore the ocean current off shore is bringing colder water southward from the Gulf of Alaska. The Gulf Stream, located off the eastern coast of the continent, is a warm current. The type of ocean current (warm or cold) influences coastal climate.

Air is warmed or cooled by the underlying surface, so cold water has a cooling effect on the overlying air. The presence of a cold ocean current offshore has a stabilizing effect on a region's climate because the cold air is more dense than the surrounding environment, and the air does not rise. The opposite occurs when a warm current is nearby. If the warmer, moist air is less dense than the surrounding environment, it will rise. Locations at the same latitude on a continent's eastern coast experience more thunderstorms during summer than locations on the western coast. The same is true for Mississippi. The temperatures of the Gulf of Mexico range from around 50°F in winter to near 90°F in summer. All that warm, moist air helps provide fuel for clouds and precipitation.

Mississippi is situated in a region where water is a bountiful natural resource, second only to Louisiana as the wettest state in the union, considering the average amount of precipitation over the state's land area. The statewide average of about 56 inches over nearly 31,000,000 acres produces a volume in excess of 142,000,000 acre-feet of water delivered to Mississippi by the atmosphere annually, providing both surface and groundwater in abundance. (An acre-foot is the volume of water that will cover 1 acre to a depth of 1 foot.) This much precipitation would cover a single football field to a depth of more than 107 million feet or more than 20,000 miles!

Water can also have a moderating effect on a place's climate, because of its great capacity to store heat. This means that the highs and lows are not as extreme as other locations further inland. The waters of Mississippi Sound modify the summer heat and winter cold. Biloxi has an average of only 55 days with a temperature 90°F or higher, while only 40 miles inland, Wiggins averages 103 such days annually. Biloxi has about 13 freeze days each year contrasted with about 40 in Wiggins.

THE NORTH AMERICAN LANDMASS

Due to differences in the way land and water gain and hold onto heat, land heats and cools more quickly than water. During summer, the land in Mississippi will quickly heat up to the 90s during the day, but the Gulf of Mexico remains in the 80s. At night, the Gulf temperature remains roughly the same, while the temperature on land falls into the 70s. Temperatures are colder in winter, but the same phenomenon can be observed. The Gulf of Mexico may be 55°F off the Mississippi coast in December, while the air temperature rises into the 60s. At night, the water will retain its heat, while the land may cool into the upper 30s. Another product of having the North American landmass to the north is the rapid swings in temperatures that are associated with mid-latitude cyclones.

Any regular weather follower can tell you that temperatures in the northern interior of the North American continent can get extremely cold during winter, while Mississippi, due to its latitude and proximity to the Gulf, stays relatively milder. This temperature contrast is greater during the cold season, and the fact that the same temperature contrast exists on a global scale is responsible for a lot of our cold season weather in Mississippi. Large temperature differences lead to differences in pressure at higher levels of the atmosphere, and these differences create faster winds. The polar jet stream (a fast-moving air current at the upper levels) exists due to temperature contrasts. The enhanced temperature difference along the jet stream is known as a front.

A front can be cold or warm with the boundary separating one air mass from another. The characteristics of an air mass are determined by whether it forms over water (a maritime air mass) or over land (a continental air mass). Continental air masses are drier. Behind a cold front air is typically colder and drier because it originated over northern North America, often Canada. The passage of a cold front during winter often brings a change from warm, humid conditions to cold, dry conditions as a continental air mass replaces a maritime air mass. The presence of the Gulf assures that the continental air mass does not stay in place for too long during winter. As the high pressure associated with the air mass moves toward the East Coast, the flow around the high changes from colder northerly winds from over land to warmer southerly winds from over the water. Every few days, this pattern repeats itself. Such is the routine in winter and spring in Mississippi.

TOPOGRAPHY OF MISSISSIPPI

On a more regional scale, topography and land cover can influence weather and climate. Mississippi has an area of a little over 47,000 square miles, with a north–south length of about 330 miles and a width of about 180 miles. The

THE PASSAGE OF A COLD FRONT

From December 3 to 5, 2008, a trough in the flow of air formed across North America. This trough extended south toward Mississippi as an area of relative low pressure formed in the southern Plains. Warm air was pulled out ahead of the front and temperatures across the state were mild for early December. Jackson saw its high temperature hit 70°F, and the Golden Triangle reported highs in the mid-60s. Overnight, the system began to pass through Mississippi, bringing with it rain and winds gusting to near 30 mph—first, warm air from the south and then cold air from the north and northwest. By daybreak in eastern Mississippi, the rain had begun to die out and big changes were on the way. Over an hour, the temperature had dropped almost 10°F and the dewpoint (a measure of the amount of moisture in the air) dropped 13°F. Between 6:00 a.m. and noon, the atmospheric pressure had increased by 0.15 inches (about 5.5 millibars), and later in the afternoon, the skies became clear and sunny. Clear conditions continued into the night, when temperatures fell into the upper 20s—much colder than the mid-50s of the previous night. While a low pressure system and passing cold front is not always followed by such a quick clearing, conditions over these few days illustrate well what happens repeatedly throughout the cold season in Mississippi.

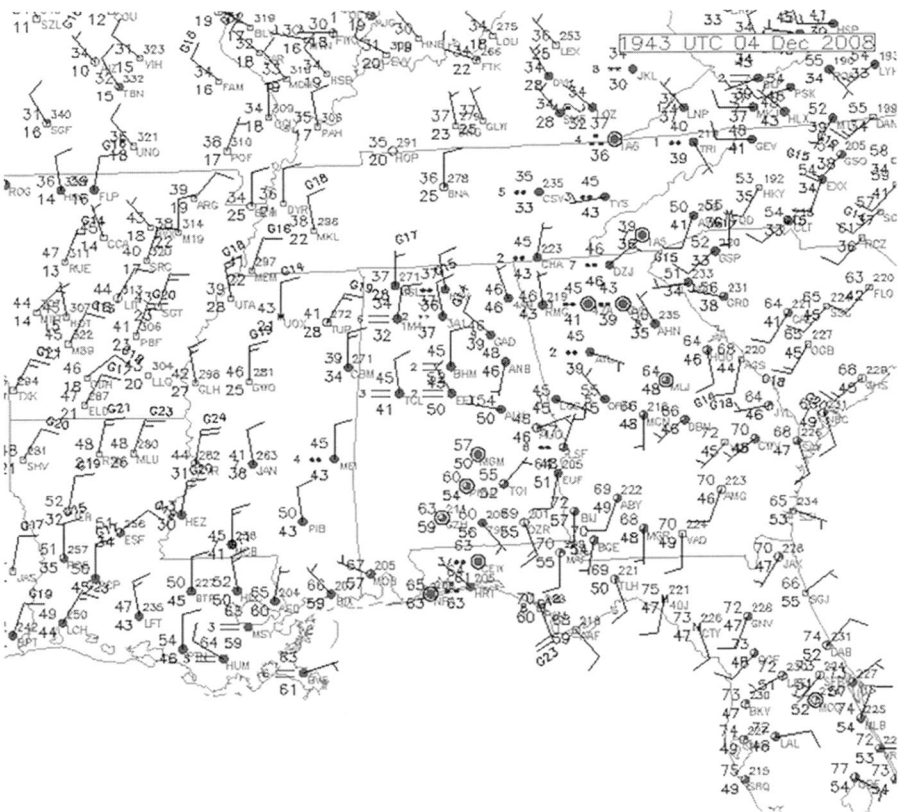

Fig. 2.3. Surface map from December 4 at 12:43 p.m. Next to each dot, the upper number is the temperature (°F) and the lower number is the dewpoint (°F). Image credit: National Oceanic and Atmospheric Administration, Department of Commerce

Table 2.1. Change in observations from the Golden Triangle Reporting Station on December 4, 2008					
Time (CST)	Wind (mph)	Conditions	Temperature (°F)	Dewpoint (°F)	Pressure (inches)
5:35 a.m.	SE 6	overcast	54	54	30.14
6:52 a.m.	SE 12	rain	54	54	30.13
7:53 a.m.	NW 14	rain	45	41	30.21
10:46 a.m.	N 15	fog/mist	39	37	30.29
1:55 p.m.	NW 12	overcast	39	36	30.33
3:50 p.m.	N 14	clear	41	30	30.36
Data from the Jackson, Mississippi, National Weather Service website					

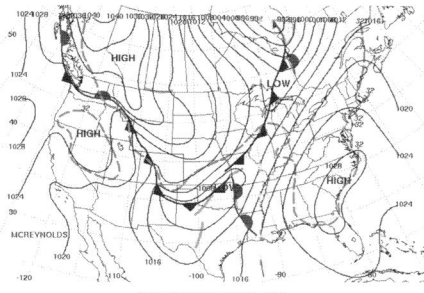

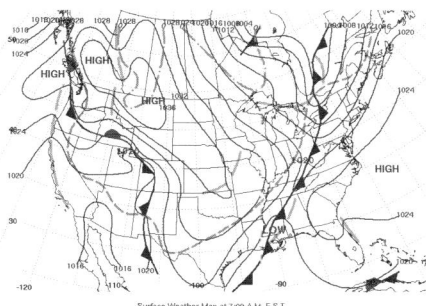

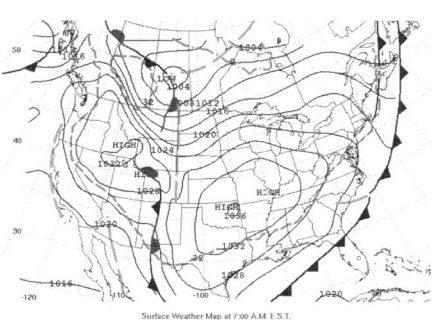

The surface map in Figure 2.3 illustrates a few of these features. If you look at Montgomery, Alabama (MGM), you can see that the front has not yet passed through. We know this because the temperature is 57°F and the dewpoint is 50°F. These are the numbers to the upper left and lower left of the station model. Wind barbs ahead of the front are indicating winds blowing from the southwest and southeast, while wind barbs in western Mississippi are blowing from the north. (Wind barbs are drawn so that they point in the same direction the wind is blowing, just as if you shot an arrow. The tip points in the direction the arrow is moving, and the feathers are closer to where the arrow came from.) The temperature in Jackson, Mississippi (JAN), is 41°F with a dewpoint of 38°F, indicating that the front has passed through Jackson. The exact position of the front at the surface is somewhere near the stations reporting light fog or mist (the two bars) or continuous rain (two dots). The best way to locate the front is to look for the area where the wind shifts and the dewpoint drops. In this case, it coincides with the area of precipitation.

Figure 2.4 shows three surface maps of the fronts and areas of high and low pressure from December 3 (before the frontal passage) through December 5 (after the frontal passage).

Fig. 2.4. Surface (from top) maps from December 3, December 4, and December 5. Image credit: National Oceanic and Atmospheric Administration, Department of Commerce

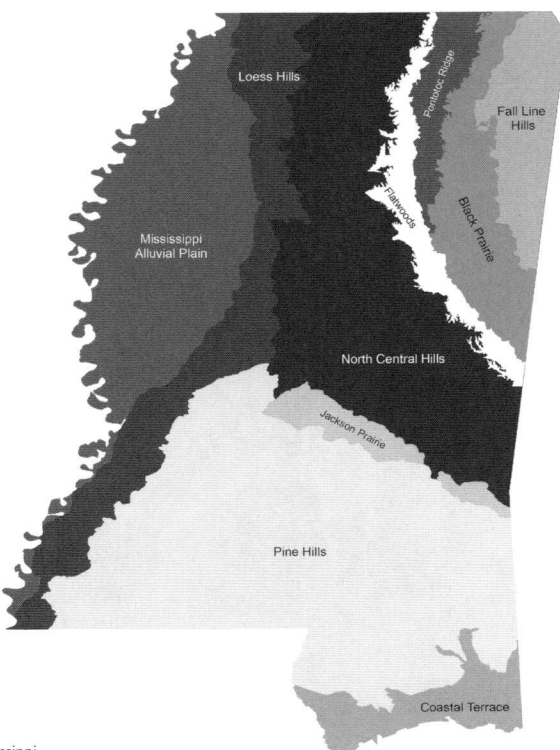

Fig. 2.5. Physiographic regions of Mississippi

state lies entirely within the North American physiographic area called the Gulf Coastal Plain Province, which is an extensive lowland bordering the Gulf of Mexico. (A province is an area of similar geology and topography.) Although the term *plain* might suggest a monotonous topography, this is not the case. Elevations range from sea level at the coast to 806 feet at Woodall Mountain in Tishomingo County in the northeastern corner of the state.

The state itself is divided into 10 distinct physiographic regions. These divisions conform closely to the geology of the state, and the soils, hydrological features, and vegetation also exhibit a generally marked relationship to the physiographic regions. Recent research has linked these landscape features to several climatic effects. For example, temperatures in the Delta, otherwise known as the Mississippi Alluvial Plain, are different from surrounding areas (Brown and Wax 2007). There is less of a temperature range in the Delta than outside of it. Also, elevation and changes in land use from farmland to forest can also influence the amount or type of precipitation an area receives. Research by Dr. Jamie Dyer (2010) at Mississippi State has indicated the transition between the open flat Delta and the slightly elevated forests of the Loess Hills may help initiate precipitation. Other features such as the Pontotoc Ridge may influence winter precipitation.

The Fall Line Hills region, also called the Tombigbee Hills, is a highland area covered with hardwood forest in the extreme northeastern part of Mississippi. Being associated with the Appalachian foothills, this area shows evidence of a long period of being carved by numerous streams. The divide between the Tennessee and Tombigbee Rivers lies in this region, but a divide cut was made to join the two systems during construction of the Tenn-Tom Waterway in the 1970s, providing a controlled water route from the Ohio Valley to the Gulf of Mexico through Mobile, Alabama.

Bordering the Fall Line Hills on the west and running from the Tennessee border south to Noxubee County is a crescent-shaped strip of level terrain known as the Black Prairie. This region is underlain by evenly eroded chalk that has weathered to produce black prairie soils. The region lacks permanent surface water features and forests and relies heavily on groundwater sources. It is used extensively for agriculture. Forming the western boundary of the Black Prairie is a narrow upland known as the Pontotoc Ridge. This feature has been deeply dissected by stream action into a series of elongated hills and valleys. It extends southward only into northern Clay County.

West of the Pontotoc Ridge is a relatively narrow and low strip known as the Flatwoods. This region is underlain by clay, which has weathered to produce a distinctive lack of relief. The region is forested, and soils are of the shrink/swell type, which is especially noticeable during periods of alternating rains and drought. These soils contribute to local flooding, which frequently occurs in late winter and spring. South and west of the Flatwoods is the large wide highland belt of North Central Hills, extending from the Tennessee border to the Alabama border. Stream action has cut deep valleys throughout this forested region, producing an area of consistently rugged relief.

On the western side of the North Central Hills and running from Tennessee on the north and Louisiana on the south are the Loess Hills. This narrow region, only 15–25 miles in width, is underlain throughout its entirety by a thick deposit of wind-blown silt called loess. Peculiar features of this region are steep cliffs lining stream courses and road cuts indicating the ability of the loess to resist ordinary erosion that wears down other types of unconsolidated sediments. Because of soil limitations, this region is no longer in agricultural use but is mainly forested. The North Central Hills gradually merge in the south into a gently rolling zone called the Jackson Prairie. This region is underlain by clays that have eroded into the characteristic subdued relief. Prairie vegetation occurs in patches surrounded by pine and hardwood forest.

Extending south from the Jackson Prairie almost to the Gulf Coast are the Pine Hills. Surface elevations in this region slope from about 500 feet on the north and west to about 100 feet at the southern margin. As the name implies, the region is covered with pine forest, which is a major contributor to the economy of this part of the state. Between the Pine Hills and the Gulf Coast lies a low flat strip of land, ranging from 5 to 30 miles in width, known as the Coastal Terrace. This strip has been deposited as coastal sediments since the Pleistocene (the geological time period from 1.8 million to roughly 10,000 years ago), and the surface descends gradually from an elevation of 50–75 feet along the landward margin to sea level at the coast. Near the coast there are considerable tracts of swamp and marsh.

On the western side of the state between the Loess Hills and the Mississippi River lies the most distinctive physiographic province in the state, the Mississippi Alluvial Plain. Locally known as the Delta or the Yazoo Basin, this region is characterized by flat, monotonous topography. Elevations range from 210 feet near the Tennessee border to 94 feet at Vicksburg, Mississippi. This general topography is further modified by extremely sinuous, shifting courses of the streams and by the presence of numerous oxbow lakes where the old course of a river has been cut off from the current flow. The alluvial soil and flat surface combine to make this a vital agricultural region. Long known for the production of cotton and soybeans, this region has more recently become the leader of catfish production in the United States, as the terrain, soils, and climate favor the establishment and maintenance of around 100,000 acres of aquacultural ponds.

Mississippi Sound, an arm of the Gulf of Mexico, forms the southern boundary of the state. In contrast to those of Louisiana, the land areas near the coastline are sharply defined, with the land rising to elevations of 10 to 20 feet behind the beaches. The coast is cut by numerous bays and a string of barrier islands parallels the coast a few miles offshore. You may wish to refer back to these regions when the different weather patterns are discussed throughout the book.

CLIMATE TYPE

Although several climate classification schemes have been proposed, the most commonly used one is the Köppen system, first proposed by Wladimir Köppen in 1918. The Köppen system classifies climate first by temperature, second by precipitation regimes, and then more finely by temperature again. Mississippi's climate type is humid subtropical, denoted as Cfa. "C" means that the average temperature of the coldest month is less than 64.4°F; this distinguishes Mississippi's climate from tropical climates, in which every month is at least 64.4°F. "f" climate types receive regular precipitation in all seasons; it comes from the German word *feucht*, which means humid or moist. This makes Mississippi different from other locations that have a distinct rainy season, such as the Mediterranean climate of southern California, which has dry summers and rainy winters. The "a" indicates that the average temperature of the warmest month is over 71.6°F and at least 4 months are over 50°F. This distinguishes Mississippi's climate from those that are more moderated by a body of water and do not get as hot during the summer; the best example is the marine West Coast climate of coastal Washington, British Columbia, and Alaska.

The humid subtropical climate type is typified by mostly mild winters without extended periods of temperatures below freezing; long, hot summers; and no routinely recurring wet or dry season. This climatic setting has given the state a traditional orientation toward agriculture and forestry.

Given the number of factors controlling climate in Mississippi, the state is sometimes characterized by a feast-or-famine situation, with the average traits seemingly never prevailing. For example, the subtropical jet stream can become active during winter, aiding the persistent development of mid-latitude cyclones in the Gulf of Mexico or in Texas. These mid-latitude cyclones can move over or near the state and bring warm, wet winter weather spells. A strong Bermuda High in summer can cause devastating drought conditions for weeks or months.

YEAR-TO-YEAR VARIATION

Mississippi's climate is also controlled to some extent by additional global mechanisms such as the El Niño and La Niña phenomena, the North Atlantic Oscillation, the Arctic Oscillation, the Pacific Decadal Oscillation, and the flow patterns associated with the Pacific North America Index. The influence of such a mechanism on weather in a different part of the world is called a teleconnection. Understanding the influence of each phase of these oscillations and index can help forecasters determine whether the upcoming season will be warmer, cooler, wetter, or drier than usual. Some also influence the occurrences of tornadoes and hurricanes in the state.

The best known of the teleconnections are El Niño and La Niña, two opposite phases of the Southern Oscillation. The Southern Oscillation refers to alternating pressure differences between weather observing stations in Tahiti and Darwin, Australia. El Niño, which is Spanish for "the boy child" was so named because it was first discovered off the coast of Peru and it appears around Christmas—a reference to the Christ child. Unusually warm waters off the Peruvian coast diminished the fishing crop, and fishermen were the first to note its seasonal recurrence. Under normal circumstances, the Trade Winds blow westward off the coast of South America and take some of the surface water with them as they go. Colder water rises up to replace the surface water. Cold water has a stabilizing effect on the air above it, so pressure over the Eastern Pacific is higher. The water transported by the Trade Winds to the Western

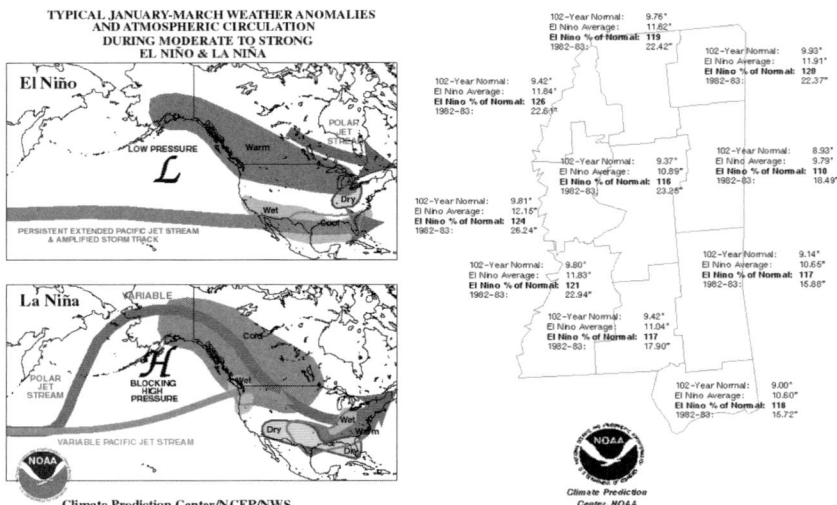

Fig. 2.6. (left) Typical El Niño/La Niña weather patterns and (right) precipitation associated with the 1982–1983 El Niño in Mississippi. For each climate division in (b) are listed the 102-year (1895–1996) average precipitation, average precipitation in an El Niño year, the percentage increase in precipitation during a usual El Niño year, and total precipitation for the 1982–1983 El Niño. Image credit: National Oceanic and Atmospheric Administration, Department of Commerce

Pacific is warmer and begins to pile up as it reaches the western side of the Pacific Basin. Warm water leads to rising air, which causes a lower pressure in the Western Pacific. During El Niño years, the pressure rises in the western Pacific and the Trade Winds blowing westward across the Pacific weaken. Without the Trade Winds pulling ocean surface water away from Peru's coast, there is less upwelling (which is what is responsible for the poor fishing during El Niño years) and the water off the coast of Peru is warmer than normal.

The opposite phase is the La Niña, which means "the girl child." In La Niña conditions, stronger than normal Trade Winds blow westward across the Pacific and enhance the upwelling of colder water off the coast of South America. The increased supply of cold water near the surface stabilizes the air near the coast and leads to drier weather (sometimes drought conditions) over Peru. Despite being phenomena of the tropics, both La Niña and El Niño influence global weather, including that in the southeastern Unites States.

Based on records from more than 100 years, El Niño events are generally associated with cooler than normal temperatures in Mississippi (by 1.5 to 2°F) and slightly wetter conditions (by 1–2 inches) than average years. During the 1982–1983 El Niño, a particularly strong event, the state received between 174% and 267% of its average amount of precipitation in November and December (Figure 2.6). One of the strongest El Niño events of the 20th century was the

1997–1998 event. While parts of the Southeast (Florida, in particular) saw record amounts of rainfall, Mississippi was near normal. The National Climatic Data Center estimated that the 1997–1998 El Niño was associated with storms, floods, and tornadoes across the Southeast that were responsible for 132 deaths and over $1 billion in damages. Even though the jet stream had the same pattern as in the 1982–1983 El Niño, most of the action it caused was just far enough east to spare Mississippi the record effects.

El Niño years are associated with a strong subtropical jet stream (Fig. 2.6a), but an otherwise zonal pattern. A zonal wind pattern is when there are no extreme ridges or troughs in the upper level wind flow. Storm systems ride along this active subtropical jet stream, but the polar jet stream shifts further north, keeping other weather systems from affecting the northern United States. The pattern associated with La Niña is considered more meridional, which means steeper ridges and troughs in the polar jet stream. El Niño and La Niña influence the number of hurricanes that make landfall, with El Niño typically suppressing hurricane activity in the Atlantic. However, it only takes one hurricane to make landfall for there to be disastrous consequences for a region. Hurricane Camille occurred during an El Niño phase (1969). There is also evidence to suggest that tornado outbreaks are more likely during the cold season in the Deep South during La Niña events.

The North Atlantic Oscillation (NAO) is similar to the Southern Oscillation in that it involves a change in pressures—this time in the Northern Atlantic Ocean. In a previous section, we discussed the Azores High, the eastern component of the Bermuda-Azores High. There is also a semi-permanent low pressure centered over the far north Atlantic called the Icelandic Low. In the two phases of the NAO, the Icelandic Low and the Azores High both strengthen (positive phase) or weaken (negative phase). When both pressures are stronger (lower pressure Icelandic Low and higher pressure Azores High), the flow tends to be more zonal; when the pressures are both weaker, the flow is more meridional. Because of this relationship to zonal or meridional flow, the positive phase has been associated with above-average temperatures in the eastern United States and the negative phase with below-average temperatures. Like El Niño, the effects are greatest during winter. The impacts of the NAO are more pronounced when the phase of the NAO and the Arctic Oscillation are both in the same phase. The Arctic Oscillation describes the change in pressure at Northern Hemisphere polar and middle latitudes. A positive Arctic Oscillation occurs when the pressure is higher at the middle latitudes (about 45°N) and lower over the poles. During a negative Arctic Oscillation, pressure is higher at the poles and lower at the middle latitudes. Like the positive phase of the NAO, the positive phase of the Arctic Oscillation helps to keep very cold air from reaching the Deep South. When the NAO and the Arctic Oscillation are both negative, winters can have long periods of cold weather.

Another teleconnection is the Pacific North American (PNA) pattern. This pattern is considered positive when a ridge is situated over western North America and negative when a trough is located over that region. The average conditions during winter are for a trough in the East and a ridge in the West. During a positive phase of the PNA, this normal pattern would be amplified. Some evidence suggests the phase of the PNA helps determine the frequency and intensity of precipitation events across the Southeast.

The Pacific Decadal Oscillation (PDO) has been associated with a change in temperature and precipitation conditions in the Southeast. The pattern is called a decadal oscillation because it usually remains in a warm or cold phase for several years or even several decades. This stability helps forecasters make seasonal predictions about temperature and precipitation. The PDO is in a warm phase when sea surface temperatures are warmer along the coast of Alaska, western Canada, and the Pacific Northwest and in a cold phase when sea surface temperatures are higher in the interior North Pacific and lower near the coast. The warm phase of the PDO has been likened to El Niño, it that it is associated with below average temperature and above average precipitation across the Southeast, including Mississippi. This influence is greatest when El Niño and the warm phase of the PDO occur at the same time.

The primary influence of these oscillations on Mississippi's weather is to help determine the strength or pattern of ridges and troughs over eastern North America. As with the relationship between the Arctic Oscillation and the NAO, these patterns can work together or work to cancel each other out. Because troughs are more potent in the Southeast in winter, this is the season during which the effects are most strongly felt. On a daily scale, ridges and troughs create Mississippi's weather by allowing the intrusion of cold air, allowing moisture to surge out ahead of a trough, preventing storm development under a ridge of high pressure, and so on. These patterns and oscillations are defined by their persistence, at least on the scale of several weeks, if not months. In terms of weather, this persistence translates to more frequent deep troughs in the Southeast or fewer weather systems making it into the Deep South. Despite their impact on large-scale conditions, there is still a large amount of daily variation in the weather.

SEASONAL VARIATION

While it may seem in the middle of August that the heat will never relent, Mississippi does have four seasons. Spring and autumn can sometimes feel like summer and winter are just alternating every few days. Because Mississippi is part of the humid subtropical climate zone, the wintertime cold temperatures are not as cold as the humid continental climates at the same longitude further north, such as Madison, Wisconsin. Also, due to the presence of the Gulf of

Mexico and lack of any barriers to its flow of moisture, Mississippi's summers do not get quite as hot as other locations at the same latitude, such as Phoenix, Arizona. Of course, Phoenix residents may quickly add that at least their heat is "a dry heat."

This section will give an overview of seasonal variation in Mississippi. Many of the topics briefly discussed in this chapter will be elaborated on in later chapters. First, though, it is important to have an understanding of why we have seasons to begin with.

All seasonal variation can be traced back to two factors: the Earth's tilt and its revolution around the sun. The earth is titled 23.5° on its axis toward the North Star, Polaris. Over the planet's long history, its tilt has ranged from 22.1° to 24.5°. The change is constant but slow—taking about 41,000 years to cycle between the two. The direction of the tilt also changes. Thuban was the North Star in 300 B.C. and Vega will become our North Star in about 12,000 years. But, for this discussion the influence of the Earth's tilt is the same as if it never moved at all. As Earth revolves around the sun, it remains titled toward Polaris. During our summertime, the northern half of the planet is pointing toward the sun. With the axis facing into the sun, the Northern Hemisphere receives the most direct of the sun's rays, those that hit at 90° to the surface. For those who are geometry inclined, if you drew a line perpendicular to the sun's rays, that line would be tangent to the Earth's surface at the latitude of the most direct rays. The tropics (the area between 23.5°N and 23.5°S) are the only location on Earth that ever receives direct 90° rays from the sun.

On the summer solstice (June 21 or 22), the sun's rays hit at a 90° angle at 23.5°N. At the equinoxes neither hemisphere is leaning into the sun, so the point that gets the most direct rays is the equator; this is true for both the vernal and autumnal equinox. Finally, at the winter solstice (December 21 or 22), the Northern Hemisphere is facing away from the sun as the Southern Hemisphere leans toward it. As it revolves around the sun, the part of the planet getting direct 90° rays from the sun slowly changes.

This change is readily apparent. Mississippi is located between 30 and 35°N. Therefore, it never receives direct 90° rays from the sun. Near the summer solstice, the sun's direct rays are much closer to Mississippi than they are on the winter solstice. The more direct the sun's rays, the greater their ability to heat the Earth's surface. This is why it is warmest during summer and coldest during winter. Because the equator gets intense solar radiation all year long, it is hot there all year long. During winter, the incoming rays from the sun hit Earth's surface at a shallower angle. This forces the same amount of solar radiation to be spread out over a larger distance and reduces the sun's ability to heat the Earth's surface, making it colder in winter.

Another related phenomenon is day length. When the Northern Hemisphere is facing into the sun during summer, the sun appears higher in the sky.

DAY LENGTH AND THE SHORTEST DAY OF THE YEAR

Table 2.2. Sunrise and sunset times at 30, 45, and 60°N			
	Pass Christian	Minneapolis	Seward
Earliest sunrise*	5:55 a.m.	5:26 a.m.	4:32 a.m.
Latest sunrise	6:54 a.m.	7:51 a.m.	10:02 a.m.
Earliest sunset	4:56 p.m.	4:32 p.m.	3:50 p.m.
Latest sunset*	8:03 p.m.	9:04 p.m.	11:27 p.m.
* Times are adjusted for daylight saving time.			

Table 2.2 shows sunrise and sunset times for Pass Christian, Mississippi (located at 30.30°N) compared with two other locations each separated by about 15° latitude: Minneapolis, Minnesota (45.07°N) and Seward, Alaska (60.18°N). Figure 2.7 shows the overall pattern: Toward the middle of the year (June), the sun rises earlier and sets later for all three locations. Next, you can see that Seward has a longer day length than Pass Christian during the summer. Minneapolis does, too, but the difference is not quite as obvious. At the beginning and end of the year, in winter, Seward's day length is the shortest and Pass Christian's is the longest. Finally, note that the range between the earliest and latest sunrise and sunset is the smallest for Pass Christian.

If we look at just the chart for Pass Christian (Figure 2.8), you may notice one additional fact. The days with the latest sunrise and earliest sunset are not the same. While the days near the winter solstice are the shortest days of the year, December 21 is not the date with the earliest sunrise or the latest sunset. The earliest sunsets occur from November 26 through December 7 in Pass Christian, and the latest sunrises take place from January 7 to 13. The first reason for this is the sun's declination, or the latitude at which the noon rays of the sun hit the surface at 90°. The declination determines how high the sun gets above the horizon and how long the sun remains in the sky. The second has to do with how we measure time. Our day is 24 hours, but the time from one solar noon (the time when the sun crosses the central meridian of the time zone, when it is at its highest point in the sky) to the next solar noon is not 24 hours. This time varies from season to season (U.S. Naval Observatory, n.d.). A solar day is longer than 24 hours around the solstices and less than 24 hours around the equinoxes. The closer to the equator you get, the further apart the times of earliest sunrise and latest sunset get. So, the days between the earliest sunset and latest sunrise are further apart in Mississippi than in Minnesota or Alaska. This is because the length of a solar day has more influence closer to the equator and the declination has a great impact closer to the poles.

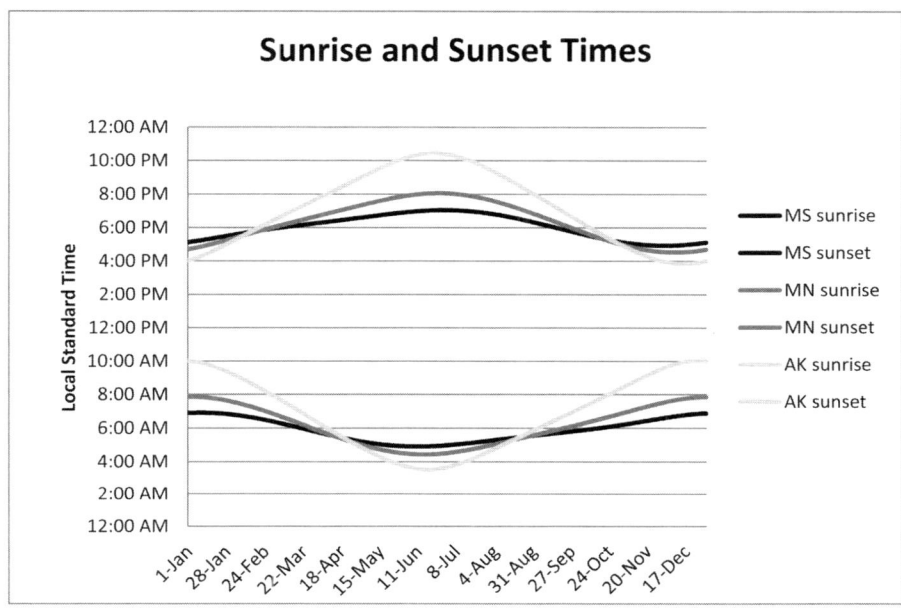

Fig. 2.7. Sunrise and sunset times for three locations at approximately 30°N (Pass Christian, Mississippi; MS), 45°N (Minneapolis, Minnesota; MN) and 60°N (Seward, Alaska; AK)

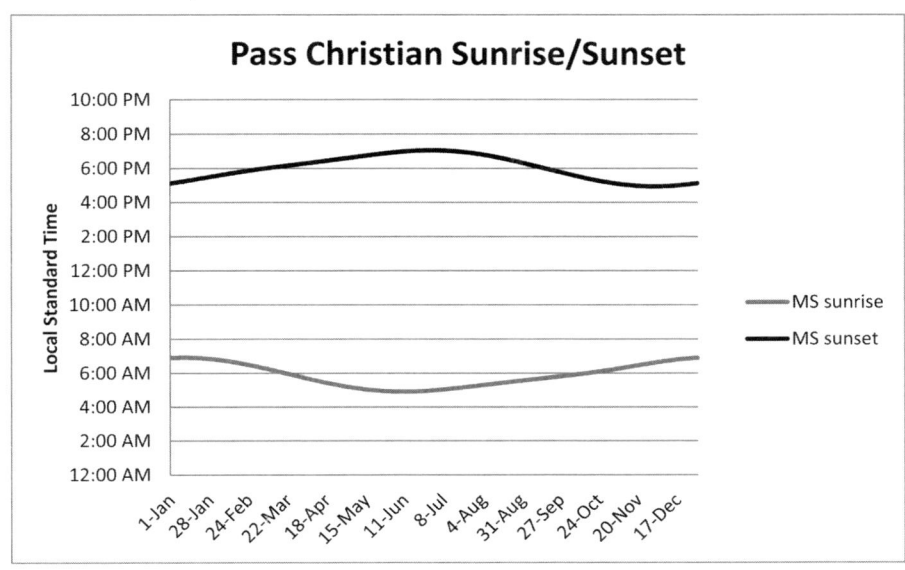

Fig. 2.8. Sunrise and sunset times in Pass Christian, Mississippi

During winter, the sun is lower in the sky. As the sun gets higher in the sky, the length of daytime gets longer. This is because more of the Northern Hemisphere is contained by the half of the Earth that is illuminated during summer than in winter. The extreme example of this can be found at the poles. On the Northern Hemisphere summer solstice, the circle of illumination contains the entire area inside the Arctic Circle (66.5°N). However, on the winter solstice, the entire northern polar region is outside of the illuminated half and experiences 24 hours of darkness. The sun rises earlier and sets later during the summer than in the winter. On June 21 in Jackson, Mississippi, sunrise is 5:54 a.m. and sunset is 8:11 p.m. On December 21, the sun rises at 6:58 a.m. and the sun sets at 5:00 p.m., providing about four fewer hours of daylight. This is perhaps most evident on those winter days when it is dark when you wake and dark again before you leave work in the evening. Further north, the days are longer in summer than in Mississippi and shorter in winter.

MISSISSIPPI CLIMATE SUMMARY

In the preceding sections, we have discussed reasons why air flows a particular way and what effect that, a body of water, the Earth's tilt, and so on has on Mississippi's climate. In this section we provide a snapshot of a year's worth of weather as defined by the state's climatology, beginning with summer.

The Warm Season
During the warm season (and throughout much of the rest of the year) prevailing southerly winds provide humid, semitropical conditions often favorable for afternoon thunderstorms. These storms produce an average of about 25% of the state's annual precipitation and are at times accompanied by locally violent and destructive winds. High humidity, combined with hot days and nights, generally produces uncomfortable conditions from May to September, with dewpoint temperatures routinely in the upper 70s. When the pressure distribution is altered so as to bring westerly or northerly circulation, periods of hotter and drier weather interrupt the prevailing humid condition. It is also not unusual for these circulation shifts to produce spells of very pleasant weather in May, June, and September, with dewpoints dropping into the 30s for a few days.

The Cold Season
In the cold season the state's weather is dominated by the positions of the polar and subtropical jet streams and their subsequent control over passages of warm and cold fronts of mid-latitude cyclones. These frontal passages alternately subject the state to warm tropical air and cold continental air, in periods of varying length. However, cold spells seldom last more than a few days. The ground

rarely freezes. When it does, it is mostly in the northern part of the state and only a few inches deep. Continental polar and Arctic air behind cold fronts is usually considerably modified by the time it enters Mississippi, but these air masses can occasionally bring large and rather sudden drops in temperature. When the upper air circulation is configured to support the intrusion of bitterly cold Arctic air masses into the state—a situation referred to as The Siberian Express—extreme cold spells may occur. Temperatures in the northern part of the state fell −19°F during one such occurrence. On the other hand, it is not unusual for warm fronts to bring air masses with temperatures in the 80s into the state during January and February.

Temperature

The normal annual temperature ranges from 61°F in the northern border counties to 67°F in the coastal counties. Daily highs in January average about 50°F in the north and about 61°F along the coast. Daily minimum temperature in January averages about 30 and 43°F in the north and along the coast, respectively. July daily average highs are about 93 and 90°F in the north and on the coast, respectively. July daily minimum temperatures average about 70°F in the north and 75°F on the coast. On average, temperatures of 90°F or higher occur just 55 days per year on the immediate Gulf Coast under the ameliorating effect of the relatively cooler Gulf waters. Inland from the coast, however, there is a rapid increase in the number of days at 90°F or higher, reaching a maximum of more than 100 days about 50 miles inland. Temperatures of 32°F or lower occur on average about 13 days a year on the immediate Gulf Coast, increasing to a maximum of about 73 days on the Tennessee border.

Temperatures exceed 100°F at one or more weather stations each summer, and 115°F has been recorded at Holly Springs. The temperature drops to 0°F or lower in Mississippi an average of once in 5 years and to 32°F or lower as far south as the Gulf Coast every winter. The lowest temperature recorded is −19°F at Corinth. Last freeze dates are quite variable, averaging from April 3 in the north to February 20 along the coast. However, one site in east-central Mississippi has had its last freeze as early as February 8 and as late as April 21.

Precipitation

Mean annual precipitation ranges from about 50 inches along the northern border to about 65 inches along the coast, averaging about 56 inches statewide. During the freeze-free season, rainfall ranges from about 24 inches in the Delta region to about 37 inches in the southeast. This distribution discourages the growth of crops with critical water requirements, such as corn, in much of the Delta and the Black Prairie provinces, but it is beneficial for cotton. Conversion from row crops to cattle in large areas in the northern part of the state is due, at least in part, to insufficient or poorly distributed rainfall. Irrigation is being

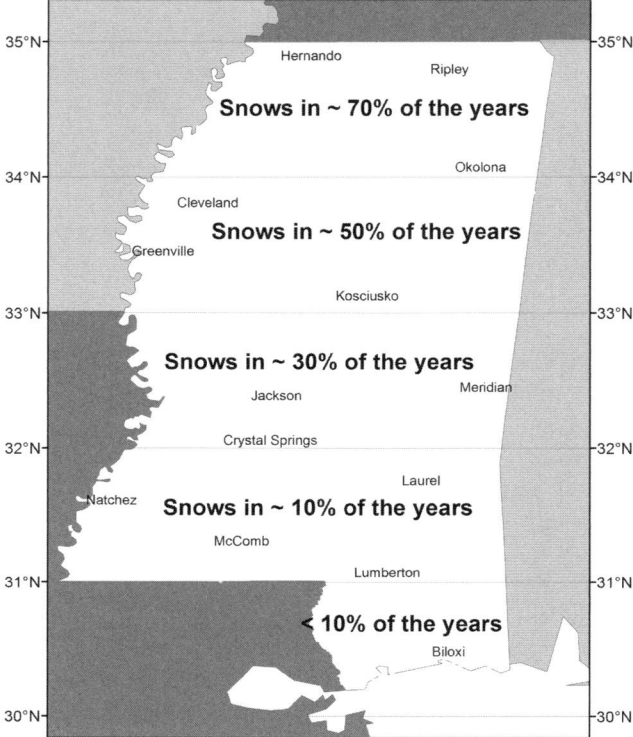

Fig. 2.9. Yearly likelihood of snowfall by latitude

increasingly practiced because the abundant rainfall does not always come in the time of greatest need. It is not unusual for Mississippi to experience general agricultural droughts, especially during summer. Stream flow and precipitation records reveal at least nine significant periods of extended drought in the state since 1930.

During winter, maximum precipitation is centered over the northern and western counties (16–18 inches) with the minimum (13 inches) on the coast. In summer, the maximum shifts to the coastal counties (19–21 inches) and the minimum to the Delta counties (9–11 inches). Spring and autumn patterns are very similar to summer. Autumn months are the driest of the year, with precipitation ranging from 8 to 13 inches, which favors the harvesting of crops. Autumn is the most agreeable season of the year, with cool nights and mild, clear, sunny conditions persisting for several days, and even weeks, at a time. The most intense rains are associated with thunderstorms. Stalled fronts and tropical storms usually cause heavier rains over longer periods of time. Daily totals may amount to more than 8 inches along the coast and more than 4 inches further inland.

While snowfall is not of much economic importance, it is not such a rare event in Mississippi as is generally believed. Measureable snow or sleet falls on some part of the state in 95% of the years. As shown in Figure 2.9, as one moves closer to the coast, the likelihood of snow in any year decreases. (For the yearly average snowfall, see Figure 7.2.) Amounts of up to 12 inches are found in the records for many individual snow events in the northern part of the state. For perspective, the record for Jackson in central Mississippi shows that 60% of snow falls in January, a snowfall greater than 1 inch occurs every 2.5 years, heavy snowfall (3 inches or greater) occurs once every 4 years, and the longest period between 1-inch snowfalls is 7 years. Ice storms occur about once every 4 years in the northern half of the state and about once every 13 years in the southeastern part of the state.

Floods

The flood season in Mississippi is from November through June (the period of greatest rainfall), with March and April being the months of greatest frequency. The season of high flows in the main Mississippi River is during the first 6 months of the year. In other streams, flooding sometimes occurs during summer from persistent thunderstorms or during late summer and early autumn from heavy rains associated with tropical storms originating in the Gulf of Mexico and passing through the state.

Local overflows occur on many streams three or four times a year in association with extended rainy spells and saturated soil conditions. Severe general flooding occurs about once every 3 years along the larger streams. The Mississippi River floods about once every 2 years from upstream runoff; the only important contribution to the Mississippi from within the state is from the Yazoo Basin. A system of levees prevents major damage from Mississippi River floods.

Severe Storms

Thunderstorms produced from afternoon heating of warm, moist air and from the passage of cold fronts and sea breezes occur on an average of 50 to 60 days a year in the northern part of the state and more than 100 days a year near the coast. Thunderstorms occur more frequently in July than in any other month, with the fewest reported from October to February. Those in late autumn, winter, and early spring produce more high winds than those in summer. However, in summer after a spell of unusually high temperatures, afternoon air mass thunderstorms may develop with local violence. During late autumn, winter, and early spring, thunderstorms may occur at any time of day, as they are usually associated with frontal activity.

Tropical cyclones (hurricanes) occur from June to November and represent a hazard to life and property in Mississippi. While these storms generally move into the state on the coast, they have on occasion entered as far north as Meridian and Greenville after crossing part of Alabama or Louisiana. The tropical cyclones are weakened quickly by passage over land, so loss of life and property due to high winds is confined mainly to the coastal areas, with losses further inland generally owing to rain damage to crops and from floods. However, the hurricane of September 26–27, 1906, which moved inland between Pascagoula and Mobile, caused great damage as far inland as Brookhaven and Waynesboro; about 10% of the virgin timber in the area was destroyed.

The hurricane of July 5–7, 1916, pursued one of the most unusual courses ever observed. It moved inland near Pascagoula late on July 5 on a northwest course as it decreased in intensity. It passed over Jackson to Cleveland, where it turned east during the night, moving over Macon to near Selma, Alabama, where a turn to the west then carried it over Birmingham and Huntsville on July 7 and 8. Another turn west took it past Nashville and into the Ohio Valley on July 10. Attending heavy rains for 3 days caused enormous losses of staple crops and resulted in great floods on the rivers of eastern Mississippi, Alabama, and Georgia.

Mississippi has been affected by high winds, high tides, and heavy rains by many tropical cyclones. The strongest was Camille on August 17, 1969, which killed 135 people in Mississippi and caused more than $5 billion (in 1969 dollars) in damages. A wind gust of 229 mph was recorded in Biloxi during this event. A more deadly hurricane struck on August 29, 2005. Hurricane Katrina made landfall in southeastern Louisiana near Buras as a category 3 storm with sustained winds of 126 mph. The storm brought nearly incomprehensible devastation and loss of life. The city of New Orleans, much of which sits below sea level, was inundated with up to 20 feet of water when several levees broke. According to Blake et al.'s (2007) *The Deadliest, Costliest, and Most Intense United States Tropical Cyclones from 1851–2006*, Hurricane Katrina caused approximately $81 billion in damages (in 2006 dollars). Uninsured or underinsured losses were estimated in the $100 billion to $150 billion range. More than 230 Mississippians and more than 1500 Louisianans lost their lives in the storm. This was the largest number of weather-related fatalities in the United States since the Lake Okeechobee storm of 1928 produced 2500 deaths.

Mississippi also commonly experiences tornadoes, many of which are violent. According to the Storm Prediction Center, 5 of the 25 deadliest ever to occur within the United States took place in the state, killing a total of 855 people (Natchez, May 1840; Purvis, April 1908; Natchez, April 1908; Starkville, April 1920; and Tupelo, April 1936). The state ranks 12th nationally in total number of reported tornadoes and 8th nationally in number of tornadoes per 10,000 square miles. A regrettable statistic is that the state ranks first nationally in tornado deaths per 1 million people. However, the tornado threat is not spatially consistent across the state. The greatest frequencies occur in Simpson, Grenada, Harrison, Humphreys, Leflore, and Jones Counties. Minimum frequencies are found on an axis from Yazoo to Itawamba Counties, in the extreme southwestern counties, and along the extreme eastern edge of the southern half of the state.

The largest number of tornadoes usually occurs in late winter and early spring (February–May), with this period accounting for about two-thirds of all tornadoes. Tornadoes do occur in all months, with the fewest occurring in the August–October period. Tornadoes in Mississippi occur at any hour of the day or night, but are least likely between 4:00 and 7:00 a.m. and most likely between 11:00 a.m. and 9:00 p.m. Nearly half of all tornadoes in Mississippi occur between 2:00 and 9:00 p.m., with the peak occurring between 6:00 and 7:00 p.m.

In short, Mississippi has a climate characterized by extreme heat in summer and by the absence of severe cold in winter. The ground rarely freezes and outdoor activities are generally favored year-round. Cold spells are usually of short duration and the growing season is long; rainfall is plentiful though not reliably distributed throughout the year. Dry spells usually accompany harvest time when they are most needed, but drought is a damaging aspect of the climate. While tornadoes and hurricanes can cause severe damage, they affect only a small part of the state at any time and protective measures can be taken against them.

Further Reading

Lutgens, F. K., and E. J. Tarbuck. 2004. *The Atmosphere: An Introduction to Meteorology*. 9th ed. Upper Saddle River, NJ: Prentice Hall.

Moran, J. M. 2006. *Weather Studies: Introduction to Atmospherics Science*. 3rd ed. Boston, MA: American Meteorological Society.

Robinson, P. J., and A. Henderson-Sellers. 1999. *Contemporary Climatology*. 2nd ed. Harlow, England: Pearson Education Limited.

Average January Maximum Temperature

- 46.9 - 49.1
- 49.2 - 51.2
- 51.3 - 53.4
- 53.5 - 55.5
- 55.6 - 57.7
- 57.8 - 59.8
- 59.9 - 62

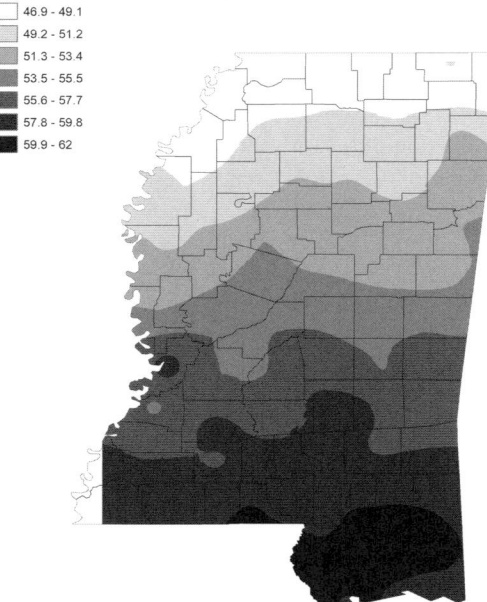

Fig. 3.1. Average maximum temperature (°F) for the month of January

Average July Maximum Temperature

- 90.3 - 90.6
- 90.7 - 90.9
- 91 - 91.2
- 91.3 - 91.5
- 91.6 - 91.8
- 91.9 - 92.1
- 92.2 - 92.4

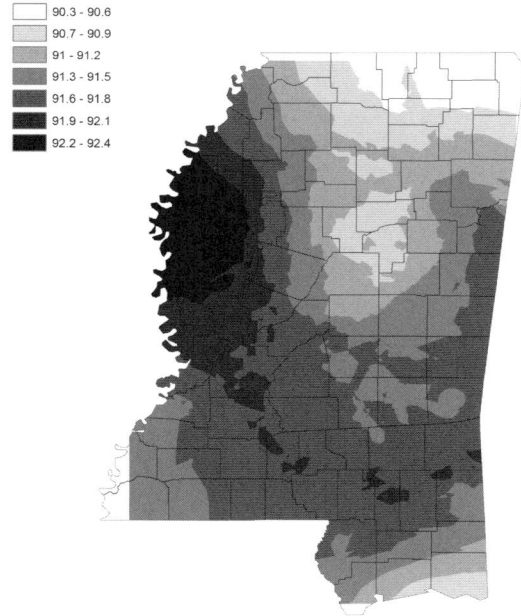

Fig. 3.2. Average maximum temperature (°F) for the month of July

3. TEMPERATURE

On July 10, 1913, the temperature at Greenland Ranch, California (in Death Valley), was 134°F. On January 20, 1954, the temperature dropped to −69.7°F at Rogers Pass, Montana. These are the highest and lowest temperatures ever officially observed in the United States. By comparison, Mississippi's highest and lowest official temperatures are 115°F on July 29, 1930, at Holly Springs and −19°F on January 30, 1966, at Corinth. Thus, while Mississippi recorded neither the hottest nor the coldest temperatures in the United States, it can still experience some very extreme temperatures. The occurrence of both the extreme records in the northern part of the state also points out the strong coastal versus continental influence on temperatures.

A map of average high temperatures in January shows a pattern one might expect (Figure 3.1). The temperature is highest in southern Mississippi, from near Hattiesburg south to the Gulf, and gradually gets cooler north to the Tennessee border. You can also see that the average high temperature for the month for January is slightly lower in the Delta than for other counties at the same latitude. On average, winter days on the Delta are colder than surrounding areas.

In summer the pattern is reversed, with cooler average temperatures in the south and warmer average temperatures in the north (Figure 3.2). This is in response to the differential heating of the Gulf to the south and the continental landmass to the north. Because water heats and cools more slowly than land, it gets out of phase with the land, moderating temperature locally. The result is warmer winter and cooler summer temperatures, known as the maritime effect. In contrast, more inland locations experience the influence of continentality, in which land heats and cools more quickly than water, providing hotter conditions in summer and colder conditions in winter. You can also see in Figure 3.2 that the Delta tends to be the warmest location in the summer.

These conditions are seen in the climate records for the coastal and northern portions of the state. For example, Corinth experiences an average of 67 freeze days each year compared to only 20 freeze days at Pascagoula. Conversely, temperatures top 90°F an average of 69 times each year in Corinth but only 54 times in Pascagoula. From another perspective, during the period 1948–2008 summer (June–August) temperatures exceeded 100°F on 87 days at Booneville in the north, but on only 30 days at Biloxi on the coast. Because Mississippians

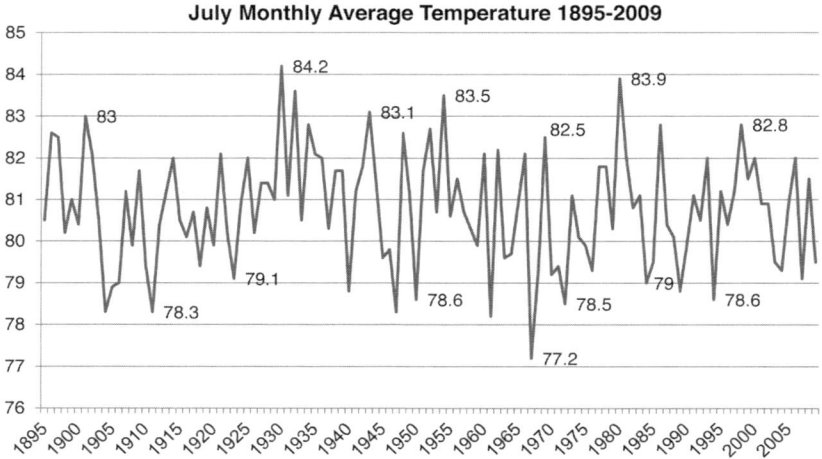

Fig. 3.3. Monthly average temperature in July, 1895–2009

are used to the heat, cold conditions probably cause more economic damage and human suffering.

Some of the hottest weather in Mississippi's history is exemplified by three July heat spells. The average monthly temperature in July is close to 81°F (Figure 3.3). July 1930 was 3.4°F above normal statewide, and it was 115°F at Holly Springs on July 29. July 1954 was almost 3°F above normal statewide, and the temperature hit 107°F at Eupora on July 16. July 1980 was 3.1°F above normal statewide, with a recorded high of 108°F in Starkville on July 17. These heat spells resulted in the three warmest Julys on record from 1895 to 2009, and the coldest July on record occurred in 1967. Figure 3.3 illustrates that even though several years may be colder or warmer than average, a particular month's temperature can vary considerably from one year to the next.

Another aspect of temperature in Mississippi is the dewpoint temperature. This is the temperature to which a mass of air must be cooled in order for condensation to occur. The dewpoint rises with the amount of moisture in the air, so a high dewpoint means a high humidity. For this reason, dewpoint temperature is a good indicator of the likelihood of clouds, fog, frost, and evaporation rates. It is also strongly related to human comfort. A good rule of thumb is that when dewpoint temperature is below 60°F, most people are comfortable. Above that threshold, the air feels muggy to most people.

Figure 3.4 shows average dewpoint temperature and average minimum temperature throughout the year at Meridian, Mississippi. Dewpoint in the summer in Mississippi is routinely in the 70s, giving the state its reputation for hot, humid summers. The figure shows that air temperature routinely cools to the

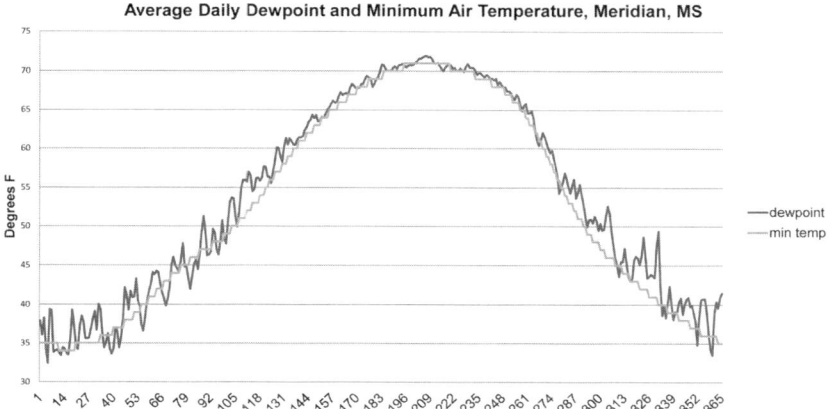

Fig. 3.4. Average daily dewpoint and minimum temperature in Meridian, Mississippi

dewpoint, especially during the warm summer days. This explains why Mississippians have dew almost every morning on grass, crops, and automobile windshields.

APPARENT TEMPERATURE: HOW HOT OR COLD DOES IT FEEL?

"It's not the heat, it's the humidity." Those words are often used to complain about the muggy feeling that comes along with summer afternoons in the Southeast, usually when the temperature here is being compared with a drier location, such as Arizona. The rationale is that, while the temperature on the thermometer might be higher in the desert Southwest, it is the Southeast that feels warmer due to the amount of moisture in the air. This phenomenon is real and is quantified with the heat index. A lot of moisture in the air can make it feel hotter than the temperature might indicate.

The heat index is used to evaluate how hot the air temperature feels to your skin, or the apparent temperature. The apparent temperature combines the effects of air temperature and relative humidity. When temperature and humidity are both high, the result is an increase in how hot the air feels. Increased humidity creates the sensation of hotter temperature by depriving the body of its greatest natural cooling mechanism, evaporation of sweat. When sweat evaporates it removes heat from the body at the rate of about 600 calories/gram of water evaporated, known as the latent heat of evaporation. When air contains high moisture levels (high humidity), evaporation is reduced. Anyone who has lived in Mississippi in the summer has experienced sweat collecting on their

NOAA NWS Heat Index

Air Temperature °F

Relative Humidity %	80	82	84	86	88	90	92	94	96	98	100	102	104	106	108	110
40	80	81	83	85	88	91	94	97	101	105	109	114	119	124	130	136
45	80	82	84	87	89	93	96	100	104	109	114	119	124	130	137	
50	81	83	85	88	91	95	99	103	108	113	118	124	131	137		
55	81	84	86	89	93	97	101	106	112	117	124	130	137			
60	82	84	88	91	95	100	105	110	116	123	129	137				
65	82	85	89	93	98	103	108	114	121	126	130					
70	83	86	90	95	100	105	112	119	126	134						
75	84	88	92	97	103	109	116	124	132							
80	84	89	94	100	106	113	121	129								
85	85	90	96	102	110	117	126	135								
90	86	91	98	105	113	122	131									
95	86	93	100	108	117	127										
100	87	95	103	112	121	132										

☐ Caution   ☐ Danger

☐ Extreme Caution   ☐ Extreme Danger

Fig. 3.5. The heat index, the apparent temperature resulting from the combined effects of temperature and humidity

skin and dripping off. In addition to being quite uncomfortable, inconvenient, and even messy, these conditions can lead to serious health consequences.

The heat index has been tabulated for assessment of the various categories of health dangers. Figure 3.5 shows how air temperature and relative humidity combine to produce the apparent temperature, as well as the attendant dangers. For example, a temperature of 94°F combined with a relative humidity of 60% causes the air to feel like it is 110°F, and places a person in a category of danger for heat disorders like heat exhaustion or heat stroke. If the temperature was 100°F with a relative humidity of 60%, the air would feel like 129°F and a person would be in extreme danger of heat-related problems with prolonged exposure or strenuous activity.

Heat is the number one weather-related killer in the United States. National Weather Service statistical data show that heat causes more fatalities per year than floods, lightning, tornadoes, and hurricanes combined. Mississippi summers are hot, and many summers see heat waves. Mississippi heat waves tend to combine both high temperature and high humidity, although some have been catastrophically dry. The heat wave of late summer and autumn 2010 saw afternoon temperatures routinely top 100°F with morning lows staying in the 70s for several weeks, along with a concurrent lack of rainfall. Environmental damage, economic loss, and human suffering result from the high heat periods in the state.

Wind Chill Chart

Calm	40	35	30	25	20	15	10	5	0	-5	-10	-15	-20	-25	-30	-35	-40	-45
5	36	31	25	19	13	7	1	-5	-11	-16	-22	-28	-34	-40	-46	-52	-57	-63
10	34	27	21	15	9	3	-4	-10	-16	-22	-28	-35	-41	-47	-53	-59	-66	-72
15	32	25	19	13	6	0	-7	-13	-19	-26	-32	-39	-45	-51	-58	-64	-71	-77
20	30	24	17	11	4	-2	-9	-15	-22	-29	-35	-42	-48	-55	-61	-68	-74	-81
25	29	23	16	9	3	-4	-11	-17	-24	-31	-37	-44	-51	-58	-64	-71	-78	-84
30	28	22	15	8	1	-5	-12	-19	-26	-33	-39	-46	-53	-60	-67	-73	-80	-87
35	28	21	14	7	0	-7	-14	-21	-27	-34	-41	-48	-55	-62	-69	-76	-82	-89
40	27	20	13	6	-1	-8	-15	-22	-29	-36	-43	-50	-57	-64	-71	-78	-84	-91
45	26	19	12	5	-2	-9	-16	-23	-30	-37	-44	-51	-58	-65	-72	-79	-86	-93
50	26	19	12	4	-3	-10	-17	-24	-31	-38	-45	-52	-60	-67	-74	-81	-88	-95
55	25	18	11	4	-3	-11	-18	-25	-32	-39	-46	-54	-61	-68	-75	-82	-89	-97
60	25	17	10	3	-4	-11	-19	-26	-33	-40	-48	-55	-62	-69	-76	-84	-91	-98

Temperature (°F)

Wind (mph)

Frostbite Times   30 minutes   10 minutes   5 minutes

$$\text{Wind Chill (°F)} = 35.74 + 0.6215T - 35.75(V^{0.16}) + 0.4275T(V^{0.16})$$

Where, T= Air Temperature (°F) V= Wind Speed (mph)

Effective 11/01/01

Fig. 3.6. The wind chill equivalent temperature resulting from the combined effects of temperature and wind speed. Image credit: National Oceanic and Atmospheric Administration, Department of Commerce

Although heat (and excessive heat) is the primary temperature concern in Mississippi's climate and weather, cold spells are also a part of the state's climate history. Several periods of extreme cold stand out. For example, the month of January 1940 was a record breaker at that time, with statewide temperature falling 14.5°F below the January normal. The lowest temperature was −14°F, recorded at Edinburg, Kosciusko, and Tupelo on January 27. It is enlightening about the nature of climate in Mississippi to note that this record cold spell occurred in the middle of a decades-long warm period from the 1920s to the 1950s.

Other notable cold spells occurred in December 1963, February 1978, January 1985, December 1989, and December 2000. A personal anecdote from one of the authors serves to explain the kinds of suffering brought to Mississippi citizens during these cold outbreaks. On December 24, 1989, 14 members of a Mississippi family were gathered for Christmas in a small three-bedroom, one-bathroom home in Booneville. When the temperature reached −6°F, all the water lines and sewer lines froze solid. There was no water, and the commode could not be flushed. These conditions persisted through the night and into Christmas Day, making the family Christmas visit of 1989 one to be remembered.

Like the heat index for hot weather, the wind chill index, or wind chill equivalent temperature, quantifies how cold temperature feels to human skin when accounting for the combined effects of temperature and wind speed. Figure 3.6

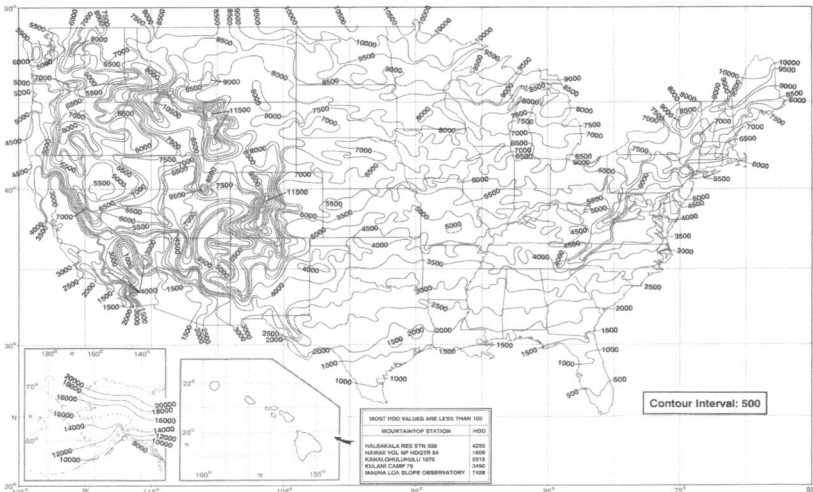

Fig. 3.7. Average heating degree days in the United States, 1961–1990. Image credit: National Climatic Data Center, National Oceanic and Atmospheric Administration, Department of Commerce

shows how these two variables combine to produce the wind chill equivalent temperature, as well as the danger categories associated with different combinations of temperature and wind speed.

The figure shows that a temperature of 10°F combined with a wind speed of 15 mph produces conditions that make the air temperature feel like –7°F and that can cause frostbite with exposure for more than 30 minutes. These conditions of temperature and wind speed are common in Mississippi in winter. If air temperature reaches 0°F with wind speeds of 20 mph, the air feels like it is –22°F and frostbite can occur with exposure of less than 30 minutes. During the Christmas 1989 cold outbreak, temperatures close to –10°F combined with wind speeds of 20–25 mph, resulting in wind chill temperatures of –37°F, at which exposed human skin can freeze in less than 10 minutes. Although these conditions are not common in Mississippi, they do occur on occasion and represent a dangerous threat to humans and animals as well as a major disruption to customary lifestyles and conveniences.

HEATING AND COOLING DEGREE DAYS

An application of temperature data known as the heating degree day (HDD) allows evaluation of the costs of heating homes and buildings. Developed by

ANNUAL COOLING DEGREE DAYS
BASED ON NORMAL PERIOD 1961-1990

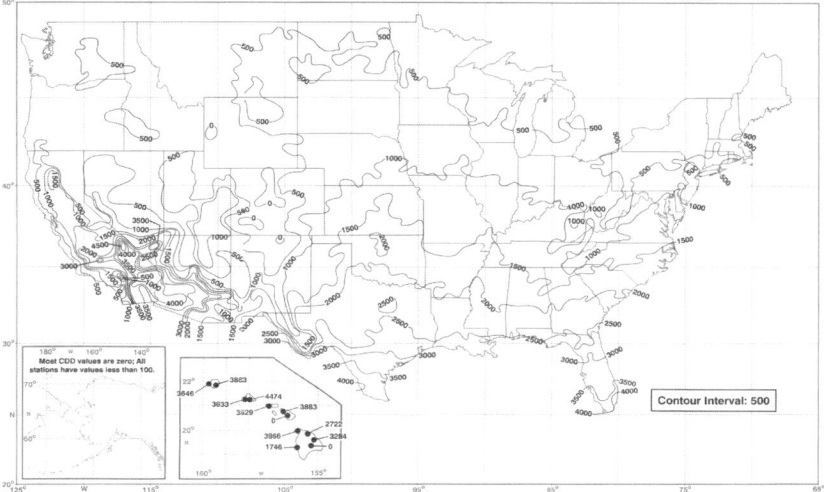

Fig. 3.8. Average cooling degree days in the United States, 1961–1990. Image credit: National Climatic Data, National Oceanic and Atmospheric Administration, Department of Commerce

heating engineers, calculating HDDs assumes that heating is not required in a structure when daily average temperature is 65°F or higher. Each degree of temperature below 65°F is counted as 1 HDD. For example, a day with an average temperature of 55°F would have 10 HDDs, and a day with an average temperature of 45°F would have 20 HDDs. The amount of heat required to keep a building at a given temperature is related to the total HDDS that accrues over a period of time, such as the winter season. The relationship is linear, meaning that doubling the HDDs means a doubling in the amount of energy, and therefore expense, required to maintain the desired temperature. This means that a month with 800 HDDs will have a heating bill twice as high as a month with 400 HDDs. When seasonal total HDDs are compared for different places, it is possible to estimate seasonal energy and heating expenses.

In this regard, Mississippi can use this application of its climate attributes to attract businesses, industry, and residents. For example, Figure 3.7 shows that Chicago, Illinois, has about 6500 HDDs in an average year, whereas Jackson, Mississippi, has an average of only about 2500. Thus, buildings in Jackson could be heated for about a third of the cost to heat the same building in Chicago. Business and industry consider such amenities of the climate when considering where to locate.

In the same manner, the costs of cooling a building can be calculated by using the cooling degree day (CDD). CDDs are accrued when average daily

Legend

Annual Heating Degree Days

- 1,518 - 1,862
- 1,863 - 2,206
- 2,207 - 2,550
- 2,551 - 2,894
- 2,895 - 3,238
- 3,239 - 3,582
- 3,583 - 3,927

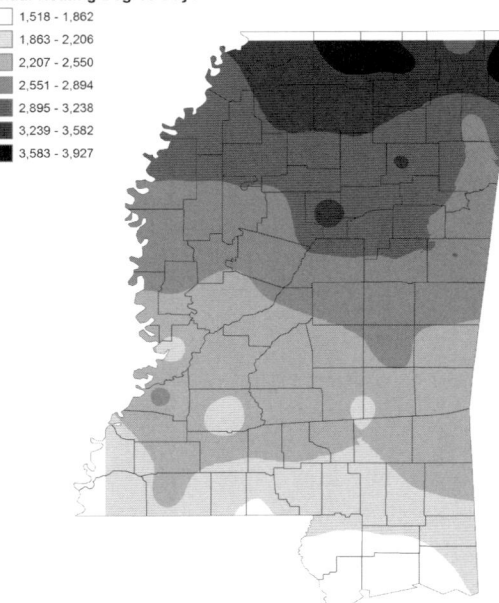

Fig. 3.9. Average annual heating degree days in Mississippi. Data from National Climatic Data Center

Legend

Annual Cooling Degree Days

- 1,328 - 1,521
- 1,522 - 1,713
- 1,714 - 1,905
- 1,906 - 2,098
- 2,099 - 2,290
- 2,291 - 2,482
- 2,483 - 2,675

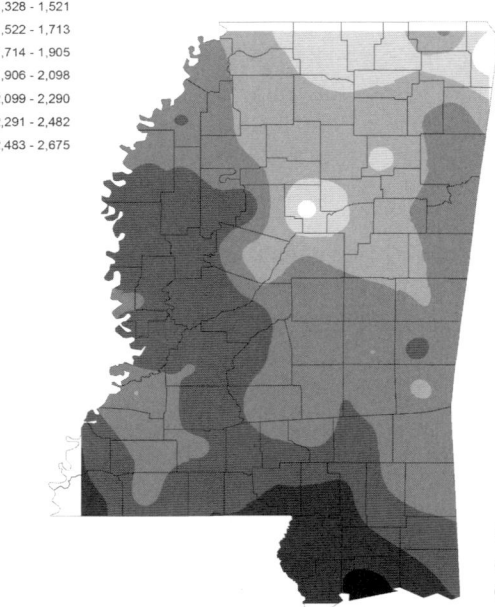

Fig. 3.10. Average annual cooling degree days in Mississippi. Data from National Climatic Data Center

temperature is above 65°F. So if average temperature on a day is 85°F, there would be 20 CDDs. Warm season totals can be used to show the comparative climate advantage of different places. For example, Figure 3.8 shows that Chicago averages only about 500–600 CDDs each year, compared to about 2000 in Jackson and 3500 in Miami, Florida.

Compared to many other areas in the United States, Mississippi has a major climatic advantage with regard to heating and cooling costs. This aspect of Mississippi's temperature regime has been used to attract new and high-paying industries to the state in recent years. Mississippi's climate provides for an annual range of HDDs from about 1500 on the coast to 3900 in the north (Figure 3.9). The range of CDDs is about 2600 on the coast to 1500 in the northern part of the state (Figure 3.10). The patterns that exist are similar to patterns in maximum temperature, with the exception of the coast. Even though the coastal counties have lower average July maximum temperatures than northern counties, they accrue a greater number of CDDs due to their longer warm season.

PUTTING DAILY TEMPERATURE OBSERVATIONS INTO THE CLIMATE PERSPECTIVE

Figures 3.1 and 3.2 were made using National Climatic Data Center monthly station normals. A normal is a climatological average taken over a period of 30 years, recalculated at the end of each decade. Normals used for this book were averages for the period 1971–2000. While climatological normals are average weather, a place's climate also takes into account the variation. It is helpful to know not only what the historical high temperature is for a particular date, but also what the typical range of temperatures and even the record high and low are for that date. Although we sometimes hear the weathercaster describe the forecast high as warmer or colder than we "should be" at that time of year, it's a little misleading. As shown in Figure 3.11, the average maximum temperature on January 1 at the Jackson, Mississippi, station is 59.3°F, but you can see there is a lot of year-to-year variation. It was 40°F on January 1, 1974, while the next year it was close to 80°F on that date. The average of that variation is slightly over 10°F. This means that it would not be that out of the ordinary for this date in January to be 10°F above or below 59.3°F.

A group of meteorologists and climatologists in Missouri found that this variation follows a normal distribution, where 68% of daily temperatures fall within a range of the average temperature plus or minus the average variation (Lupo et al. 2003). A normal distribution is a frequency distribution that assumes the familiar bell-shaped curve (Figure 3.12). The data graphed are daily differences, or deviations, between recorded high temperature and the average for that date. Figure 3.12 illustrates that over a long enough period of time, most daily high temperatures will be somewhat close to the average (that is, a

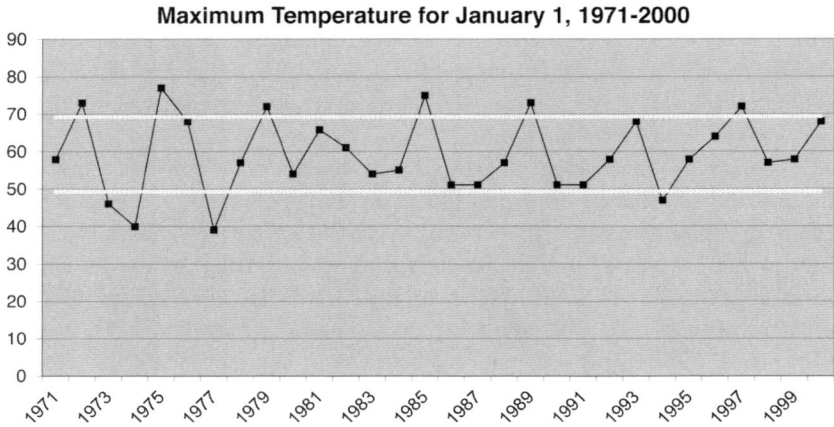

Fig. 3.11. Yearly variation in high temperature on January 1 (1971–2000) in Jackson, Mississippi. Light colored lines show the average range of variation

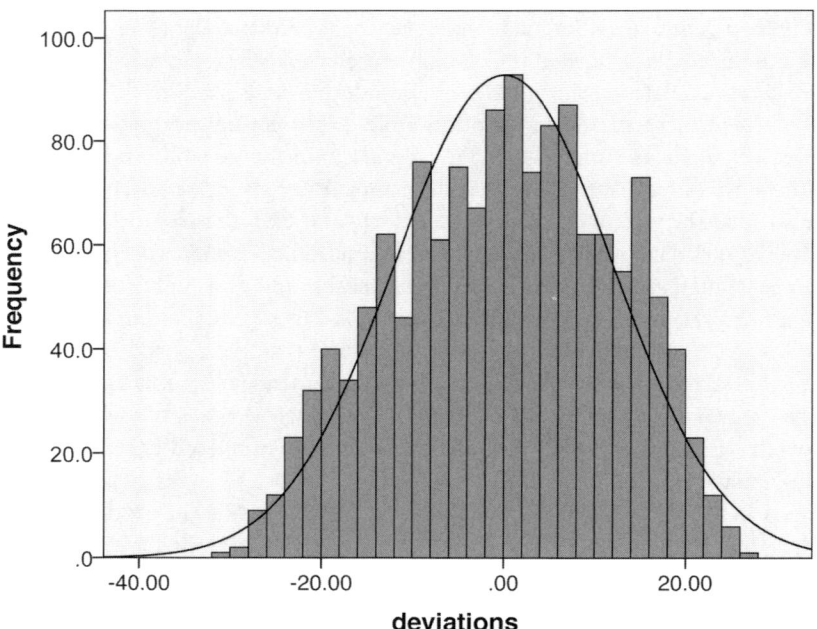

Fig. 3.12. Deviation between January daily high temperatures and January average high temperatures from 1971 to 2000 in Jackson, Mississippi, with a normal curve overlaid

deviation of zero), but over time, there will also be a number of high tempera-tures that are different from the average by 20°F or more. The fact that these temperatures are normally distributed lets us know how common it is for a temperature to be 10, 20, or even 30°F above the average (likely, not very likely, and fairly unlikely, respectively). A follow-up study by meteorologists in North Carolina also found that the amount of variation between average temperatures and actual temperatures has a seasonal component (Holder et al. 2006). This is also true in Mississippi. The average variation in January high temperatures is 12°F, but the average variation in July is only 4°F. This means that the actual high temperature will typically be closer to the normal high in July than in January.

Further Reading

To compare Jackson, Meridian, or Tupelo degree days with other locations around the country, check out the following link from the National Climatic Data Center: http://www.ncdc.noaa.gov/oa/climate/online/ccd/nrmhdd.html

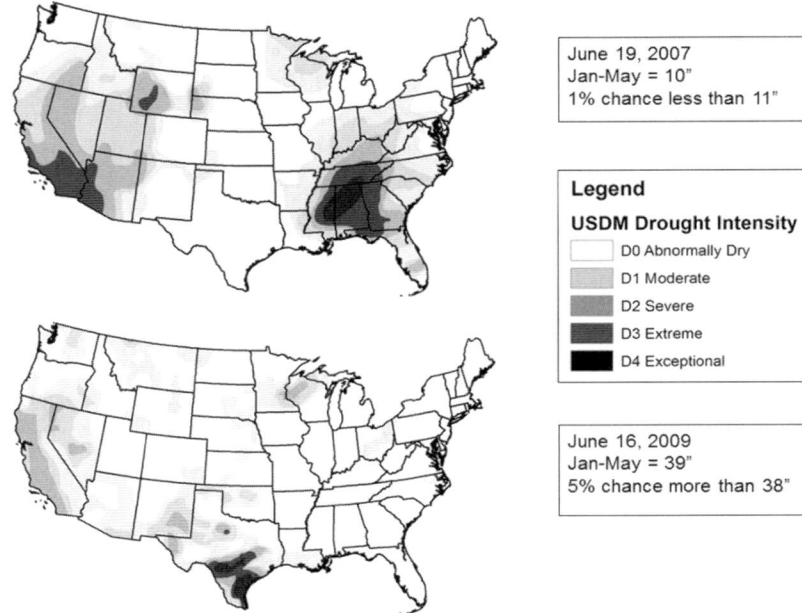

June 19, 2007
Jan-May = 10"
1% chance less than 11"

Legend

USDM Drought Intensity

D0 Abnormally Dry

D1 Moderate

D2 Severe

D3 Extreme

D4 Exceptional

June 16, 2009
Jan-May = 39"
5% chance more than 38"

Fig. 4.1. Rainfall and drought distribution illustrate the "feast or famine" nature of Mississippi's precipitation regime. Maps from U.S. Drought Monitor

4. PRECIPITATION: FEAST OR FAMINE

The basic source of Mississippi's water resources is precipitation, which delivers moisture to the state. Mississippi is situated in a region where water is an abundant natural resource, with annual average precipitation ranging from about 51 inches in the north to about 64 inches on the coast, and a statewide average of about 56 inches. That statewide average over the state's 30,538,240 acres produces a volume of 142,002,810 acre-feet of water delivered to the state by the atmosphere each year. This is a renewable natural resource of impressive dimension and extreme importance. If 56 inches of precipitation fell over the entire state every year, and if it was evenly distributed through the year, Mississippi would be a far different place.

The reality is that the distribution of the resource is highly variable in both time and in space. One year may be exceedingly wet and be followed by a record dry year. Also, one part of the state may be experiencing wet conditions, while other parts of the state are suffering drought. The precipitation regime that creates the water resources therefore provides both opportunities and constraints for human use of the environment. Adjustment to these climatic limitations has led to the present arrangement of agricultural and forestry activities, urban and industrial water supply types and attributes, and other land use patterns and resource use considerations.

Figure 4.1 illustrates the "feast or famine" nature of the precipitation regime in Mississippi. When the growing season began in 2007, the east-central part of the state had received only 10 inches of precipitation since January 1. Climatologically, there is only a 1% chance of receiving less than 11 inches of precipitation between January and May. That led to the extreme drought conditions seen in the figure in June 2007. Two years later, east-central Mississippi had received 39 inches of precipitation when the growing season began, and there is only a 5% chance of receiving more than 38 inches from January to May. That led to a completely different water situation in the state. In 2007 crops could not be planted because it was too dry, and in 2009 crops could not be planted because it was too wet. Both situations cost the state millions of dollars in lost agricultural productivity.

Precipitation is recorded in calendar years, but the impacts of it are better measured in terms of the water year, which is more commonly used by those

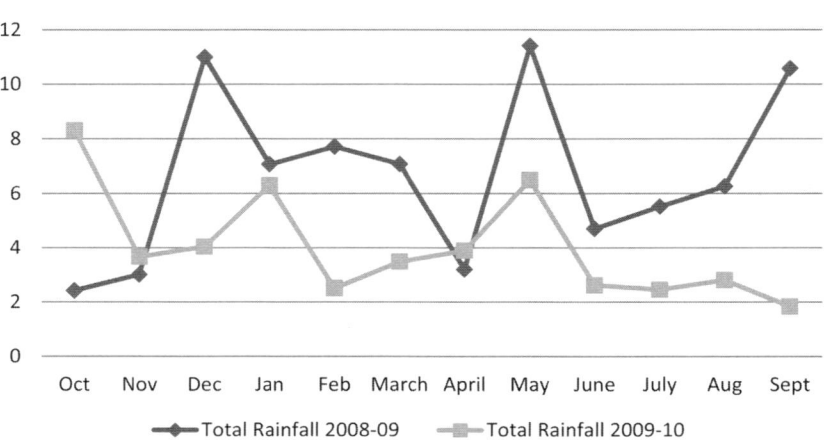

Fig. 4.2. Monthly rainfall for 2009 and 2010 at the State University observation station on the Mississippi State University campus

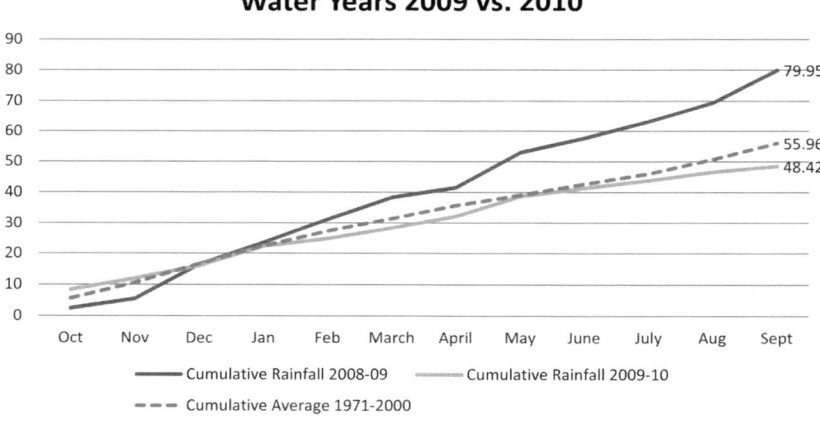

Fig. 4.3. Cumulative distribution of precipitation, 2009, 2010, and 30-year average (1971–2000) at the State University observation station on the Mississippi State University campus

who study what happens to the rain once it hits the ground. A water year begins in October and ends in September. Autumn is a dry time in most places, so the water year begins in the autumn before any snowfall-related melting and runoff. The water year also keeps the flood season together in one year. For example, the year that ends September 2010 would be considered the 2010 water year.

Figure 4.2 also illustrates the highly variable nature of precipitation in the state. The 2009 water year was an extremely wet year, setting new annual total

Legend

Average Annual Precipitation (Inches)

- 51.1 - 53.8
- 53.9 - 56.5
- 56.6 - 59.2
- 59.3 - 62
- 62.1 - 64.7
- 64.8 - 67.4
- 67.5 - 70.1

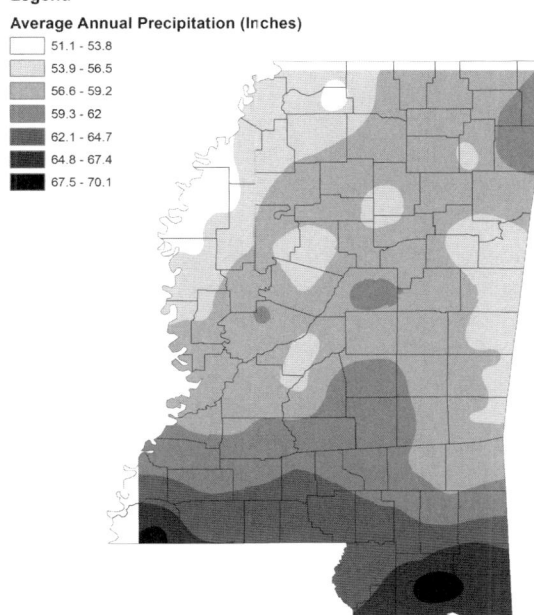

Fig. 4.4. Average annual precipitation. Data from National Climatic Data Center

records in several places in the state, whereas the 2010 water year was, with a few exceptions, a dry year characterized by continuous drought at the end of the year. Figure 4.3, illustrating the cumulative distribution of precipitation through those 2 years and a 30-year average, shows that precipitation in the 2009 water year started below average and then went well above average for the remainder of the year, ending with almost 80 inches of precipitation. In contrast, the 2010 water year started above average and then declined and re-mained below average for the rest of the year, ending with only about 48 inches of total precipitation for the year. Once again, this points out the variable nature of precipitation in Mississippi and lends credence to the idea of feast or famine in relation to Mississippi's climatic characteristics.

RAINFALL CLIMATOLOGY

Figure 4.4 shows the average annual precipitation received throughout the state. The greatest amount of rain falls along the coastal counties, which receive almost 65 to 70 inches of rainfall in an average year. The coast receives the bulk of the state's rain during the summer, while the northern and western parts of the state see the greatest amount of rain during winter. The counties of the

Delta receive the least average annual rainfall statewide. This is still more than 50 inches on average, but the fact that only about half of it occurs during the frost-free season limits the type of crops that are grown commercially.

FLOODS: THE FEAST

Rainfall in Mississippi often comes when it is least needed, and there have been many instances where an abundance of rainfall led to flooding. One of the most culturally significant floods was the Mississippi River flood of 1927. The flood of 1927 can trace its origins back to late summer 1926 (Henry 1927). A rainy pattern took hold in August from Kansas through the Ohio Valley and continued through October. With saturated soils and tributary rivers higher than normal, the stage was set for a significant flood should heavy rain occur again in the spring. In the *Monthly Weather Review*, H. C. Frankenfield and colleagues (1927, p. 16) summarized the situation, "The foundation was so well laid, that there was needed neither a prophetic vision nor vivid imagination to picture a great flood in the Lower Mississippi River in the following spring, contingent only upon a rainfall substantially above the normal quantity during the winter months."

As winter progressed into spring, many locations on the Mississippi and rivers throughout the Mississippi drainage basin spent days or weeks at flood stage. Beardstown, Illinois, on the Illinois River (a tributary to the Mississippi) spent 186 days at or above flood stage. The Ohio River surged above flood stage several times, for at least a week at a time from Evansville, Indiana, to where the Ohio meets the Mississippi near Cairo, Illinois. March 1927 was especially rainy in the Mississippi Delta. In the book *Rising Tide*, author John Barry (1997) shares the following diary entries of Greenville resident Henry Waring Ball:

MARCH 7:   Rainy
MARCH 8:   Pouring rain almost constantly for 24 hours
MARCH 9:   Rain almost all night
MARCH 12:   After a very stormy day yesterday, it began to pour torrents about sunset, and rained very hearty until 10 . . . [At] daylight, a steady, unrelenting flood came down for four hours. I don't believe I ever saw so much rain.
MARCH 18:   A tremendous storm of rain, thunder and lightning last night, followed by a tearing wind all night. . . . Today is dark, rainy and cold with a gale blowing.
MARCH 19:   Rain all day
MARCH 20:   Still raining hard tonight
MARCH 21:   Quite cold. Torrent of rain last night.
MARCH 26:   Bad. Cold rain.
MARCH 27:   Still cold and showery
MARCH 29:   Very dark and rainy

Fig. 4.5. Water on both sides of the levee in Natchez, Mississippi, during the flood of 1927. Photo credit: Family of Captain Jack Sammons, National Oceanic and Atmospheric Administration, Department of Commerce

MARCH 30: Too dark and rainy to do anything
APRIL 1:   Violent storm almost all night. Torrential rains, thunder, lightning, high winds
APRIL 5:   Much rain last night
APRIL 6:   Rain last night of course.
APRIL 8:   At 12 it commenced to rain hard. I have seldom seen a more incessant and heavy down-pour until the present moment. I have observed that the river is high and it is always raining . . . we have heavy showers and torrential downpours almost every day and night. . . . The water is now at the top of the levee.

Rainfall reports made by *Monthly Weather Review* confirm that a substantial amount of rain fell over Greenville, as well as the rest of the Mississippi drainage basin during that time. Of course, the rain did not end on April 8. Between March 5 and April 22, Greenville received 22.9 inches of rain—almost half the yearly average. The persistence of the rain must have seemed biblical. Until Easter weekend of 1927, the levees were holding.

The 19th century had seen the rapid creation of levees along the Mississippi River tasked with both improving navigation and controlling floodwaters. By 1858, more than 1000 miles of levees marshaled the river, and the Mississippi River Commission was created in 1879 to oversee the development of the levee system (Public Broadcasting Corporation). The first report to congress by the Mississippi River Commission in 1880 acknowledged that the levee system was in its "most perfect condition" between 1850 and 1858 and that it had been interrupted many times since. Much of the history of attempts to control the

Fig. 4.6. Extent of the flooding in 1927 in Greenville, Mississippi. The river stage was at 46.8 feet. Photo credit: Steve Nicklas, National Oceanic and Atmospheric Administration, Department of Commerce

Mississippi River cycled from destructive floods to acts or policies to make the levees more protective. Congress reacted to floods in 1897, 1903, 1913, and 1922 by issuing reports stating what went wrong or through acts to attempt to control the flow. By 1926, the Mississippi River Commission proclaimed that the levee system was "now in condition to prevent the destructive effects of floods."

Through the early spring rains of 1927, this remained true. Farm hands, African Americans, and convicts were forced into work gangs to shore up the Washington County levees. Then, on April 15, 1927, Good Friday, the sky dumped more than 8 inches of rain on Greenville. Elsewhere along the Mississippi, stations reported rainfall as high as 6 to 15 inches. The next day, the flood began as 1200 feet of levee collapsed near Cairo, Illinois. Tensions grew in the Lower Mississippi Valley along with intense efforts to protect and fortify the levee system. Less than a week later, the destruction would move south to Mississippi. At 7:00 a.m. on April 21 at Mounds Landing, Mississippi, the levee broke through a crevasse, flooding 2000 square miles and inundating the city of Greenville. A *Time Magazine* article from May 2, 1927, described the river as "a surly brown earth serpent uncoiling [as] the great river straightened its devious winding down a crow's-flight line of 600 miles."

Many livestock drowned, and many residents were evacuated by boats to refugee camps, which were called "concentration camps" at the time, according

Fig. 4.7. A refugee camp at Vicksburg, Mississippi, during the flood of 1927. Photo credit: Steve Nicklas, National Oceanic and Atmospheric Administration, Department of Commerce

to R. T. Lindley in a report on the flood (Frankenfield et al. 1927). Human trag-edy and controversy followed as White residents were evacuated, while many local African Americans were first stranded on the levees and then forced to work—even at gunpoint—in the Red Cross relief camps or the clean-up efforts.

The Mississippi River flooding led to the Flood Control Act of 1928 and helped Herbert Hoover, "The Great Humanitarian," become elected president based on good publicity he received directing the relief efforts. After the wa-ters receded, however, another flood continued. African Americans had begun to leave the South in large numbers during World War I, and record numbers left during the 1920s. John Barry (1997, p. 417) wrote, "The great flood of 1927 was hardly the only reason for blacks to abandon their homes. But for tens of thousands of blacks in the Delta of the Mississippi River, the flood was the final reason."

Lindley's report noted that the Mississippi River was at flood stage in Vicks-burg for 163 days between January and July 1927 (Frankenfield et al. 1927). Also notable were the Tallahatchie River at Swan Lake, which was at flood stage for 158 days from December to June, and the Yazoo River at Yazoo City, at flood stage for 185 days from January to July. The flood stage record set on May 4, 1927, remained in effect in Vicksburg until it was broken by the flooding in May 2011 (Figure 4.8). The 2011 flood also surpassed the previous record flood stage in Natchez, set in 1937.

Fig. 4.8. The former Yazoo and Mississippi Valley Railroad Depot Building in Vicksburg, which was undergoing renovation, was surrounded by floodwater in May 2011. Photo credit: K. Sherman-Morris

Widespread flooding occurred in May 2011 along rivers throughout the Lower Mississippi River Valley, including the Mississippi, Yazoo, and Big Black Rivers (Figure 4.9), primarily in areas not protected by levees. Major flooding was reported from the southern tip of Illinois to Louisiana, where the Morganza Floodway, located north of Baton Rouge, was opened to relieve flooding concerns for Baton Rouge and New Orleans. Protection for these urban areas came at the expense of rural areas to the south of the floodway.

When thinking about Mississippi River floods, the widespread 1993 flooding may come to mind. While this was a major disaster for the United States, the Mississippi River flood of 1993 was not a significant event for the state of Mississippi. The flooding was most severe in Missouri and Illinois, on the Mississippi River and its tributaries further upstream. The 1927 flood was more significant in most measurable ways except for financial loss. The river volume and area flooded were greater, more lives were lost (246 people in 1927 vs. 47 in 1993), far more people were displaced (700,000 vs. 74,000), and a greater number of buildings were damaged or destroyed in 1927.

Residents of Mississippi's capital city may also remember the events that took place Easter weekend of 1979. Like 1926–1927, the winter of 1978–1979 saw above-average rainfall. In central Mississippi, the areas near the Pearl River received 300% of the normal amount of rain. April also began rainy, with 3 to

Fig. 4.9. Flooding closed Highway 61 at the Big Black River between Vicksburg and Port Gibson in May 2011. Photo credit: John Morris

7 inches of rain falling in the first 9 days. By April 11, a low pressure system formed in Colorado and warm, moist air began to rise from the Gulf, becoming a warm front. A boundary had set up in central Texas with drier air in the Texas Panhandle and moister air in eastern Texas. Temperatures would rise into the 70s throughout much of Texas later in the day. A strong jet stream was digging into the Southwest, and conditions were coming together for severe weather. Thunderstorms developed along the boundary between the moist and the dry air and continued through the night, producing more than 30 tornadoes. One tornado was the infamous Wichita Falls tornado, which caused 42 fatalities and hundreds of millions of dollars in damage.

By the next day, the warm front had passed over Mississippi and the whole low pressure system was moving northeast, pulling a cold front along with it. As the system moved through the Southeast, 15 tornadoes were spawned in Arkansas on April 11 and one formed in Coahoma County in Mississippi. More notable for central Mississippi was the rain. A squall line ahead of a cold front moved into Mississippi on April 11, bringing with it heavy rainfall over mostly western Mississippi. Jackson recorded more than 5 inches, including 4 inches reported in 1 hour alone. The next day the front stalled again, causing heavy rain and thunderstorms.

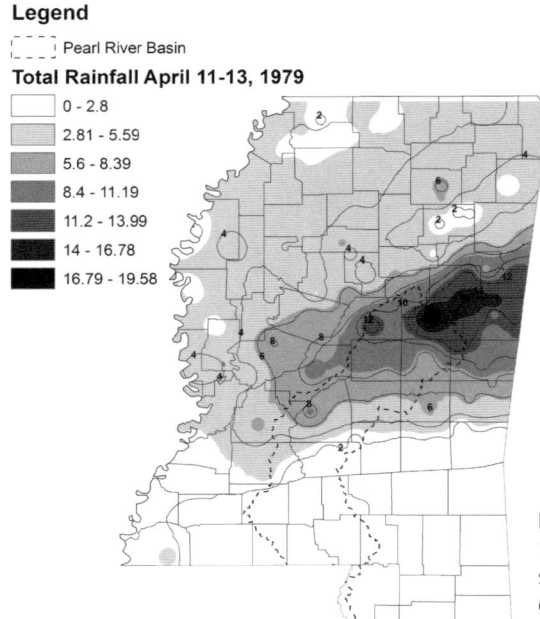

Legend

Pearl River Basin

Total Rainfall April 11-13, 1979

0 - 2.8

2.81 - 5.59

5.6 - 8.39

8.4 - 11.19

11.2 - 13.99

14 - 16.78

16.79 - 19.58

Fig. 4.10. Rainfall totals reported April 11–13, 1979. Precipitation values (inches) are shaded in, with selected values denoted by contours. Data from National Climatic Data Center

The precipitation-producing mechanisms in the atmosphere over Mississippi did their job well. Warm, moist air from the Gulf of Mexico was lifted violently to condense into line after line of heavy, slow-moving thunderstorms that developed and redeveloped over many of the same areas. Columbus and Jackson reported more than 8 inches of rain during those 2 days, with Columbus receiving more than 5 inches on April 12. Additional rain fell on April 13, as the front finally pushed through the state. Although the heavy rain ended early in the day, the trouble was just beginning. The stalled front essentially mirrored the north–south orientation of the Pearl River Basin; pouring inch after inch of rainfall inside the drainage divides of the basin and concentrating runoff into the Pearl River.

Further upstream, in the headwaters of the Pearl River, Louisville, Mississippi, reported nearly 20 inches on those 3 days, contributing to just over 24 inches for April—the most recorded for 1 month in Louisville in the 20th century. A report by the U.S. Geological Survey and the National Oceanic and Atmospheric Administration stated that the location with the greatest amount of rain was located 14 miles east-southeast of Louisville and saw 21.5 inches of rain in only 32 hours (Edelen et al. 1986). This amount of rainfall resulted in an estimated 16–20 inches of standing water over about 100 square miles in the upper part of the Pearl Basin. Figure 4.10 shows the core of extreme rainfall amounts

Table 4.1. Rainfall totals for April 12–13, 1979	
City	Rainfall (inches)
Louisville	19.6
Bluff Lake	16.3
Crawford	15.5
Kosciusko	13.0
Edinburg	10.0
Ackerman	9.0
Canton	8.3

during this storm system in central Mississippi, and Table 4.1 lists rainfall totals for some of the worst-hit cities.

These meteorological events and the resulting extraordinary amounts of rainfall ultimately produced the flood in Jackson, but the flood's severity was actually a consequence of other coinciding events—the simultaneous occurrence of patterns that overlapped because of timing and location. For example, antecedent conditions of rainfall and soil moisture storage coupled with the seasonal low energy demand for evaporation and transpiration created the potential for a high percentage of any precipitation that occurred to be converted to surplus, available almost immediately for surface runoff. Furthermore, as noted above, the drainage divide of the Upper Pearl River Basin was aligned with the pattern of rainfall over the state, so as to funnel an unusually disproportionate amount of the state's distribution of rainfall into the Pearl River during those 2 days.

Severe flash flooding had already occurred in Jackson, along with tree branches and power lines being knocked down by storms on April 11. Now, with so much rainfall over the Pearl River Basin, advisories were issued for the Pearl River. The National Weather Service warned that flooding was likely, and the mayor of Jackson, Dale Danks Jr., issued a state of emergency on Friday April 13. The Pearl River, protected to 42 feet by levees, was expected to crest at 40 feet. The skies were fair for the remainder of Easter weekend, but the Pearl continued to rise until it reached its crest on the afternoon of April 17, at 43.28 feet.

Stories reported by the *Clarion-Ledger/Jackson Daily News* chronicled what happened next (reprinted in Hederman 1979). On Friday afternoon, hundreds of people began to evacuate their houses. The River Road area, parts of Rollingwood and Westbrook Road, Sedgwick Drive, the trailer parks along I-55, Stoke Robertson Road, and parts of South West Street flooded. Thousands more were evacuated on Saturday. Floodwaters cut off Lakeland Drive from the Interstate by afternoon, and I-55 South was closed around 5:00 p.m. between Daniel Lake

Fig. 4.11. Discharge from the Ross Barnett Reservoir under Spillway Road. Photo credit: Jackson National Weather Service, photographer unknown, National Oceanic and Atmospheric Administration, Department of Commerce

Boulevard and Savannah Street. By Sunday, parts of Flowood, Richland, and downtown Jackson flooded as "flood fighters" (p. 41), including 200 prisoners from Parchman and the Hinds County Penal farm, filled sandbags to reinforce the levee. Floodwaters began to overtop the Jackson levees on Sunday and continued on Monday. The low-lying area surrounding the fairgrounds was underwater, and downtown Jackson was "paralyzed by floodwaters" (p. 47). Workers continued to reinforce the eastern levees protecting Pearl and Flowood. By Tuesday afternoon, the river at Jackson had begun to drop.

The eastern levees held, but the overall toll from the flooding was great. The U.S. Army Corps of Engineers Vicksburg District reported 1935 houses and 775 businesses flooded. According to the Mississippi Emergency Management Agency, 15,000 people evacuated. Several sources estimated damages at $500 million, and federal assistance totaled $375 million. In addition, the damages were not contained in the Jackson area. Further downstream, flooding was reported in Georgetown and Columbia.

Mississippi Governor Haley Barbour proclaimed April 12–18, 2009, as "Easter Flood of 1979, 30th Anniversary Commemoration Week." Unfortunately, little has been done to address the underlying issue of development on the flood plain of the Pearl River, so the possibility exists of another damaging flood of the same magnitude in the Jackson area.

There have been numerous other floods in the state with different causes and different outcomes. For example, the historic flood and high water stages on the Mississippi River in 1973 actually caused the Yazoo River and its tributaries to flow backward at times during the spring of that year. While traveling across the Delta during that time, it was not uncommon to see houses ringed with sandbag levees that looked like islands out in the middle of flooded fields. Also, extremely heavy late winter and early spring rains in 1991 caused flooding along the Tombigbee River. Newly developed recreational homes in Bigbee

Fig. 4.12. Flooding in downtown Jackson during the 1979 Easter flood. Photo credit: Jackson National Weather Service, photographer unknown, National Oceanic and Atmospheric Administration, Department of Commerce

Fig. 4.13. Discharge from Ross Barnett Reservoir under normal conditions. Spillway Road has been widened to four lanes since the 1979 flood to accommodate the great amount of development that has occurred in the surrounding area. Photo credit: K. Sherman-Morris

Valley were underwater. Much of the subsequent construction in those areas has been on raised stilts. Thus, the blessing of plentiful rainfall brought by Mississippi's climate can be too much of good thing when it comes under certain circumstances.

GETTING TECHNICAL: PROCESSES ASSOCIATED WITH PRECIPITATION

The sweat that glistens on a glass of sweet tea, the steam that rises from the road after a warm rain, and the moisture that forms on the grass after a cool night are all the result of the same process: condensation. Condensation is what normally occurs when water goes from the gas to the liquid state. Air naturally has a small amount of water present as a gas. This water vapor comprises no more than 4% of the atmosphere by volume, but it has very important consequences for our weather.

When discussing the amount of water vapor in the air, it is more common to talk about it as humidity. When the meteorologist on television talks about humidity or when people talk about it being humid, they are most likely talking about relative humidity. The *relative* in the name is relative to the amount of moisture air has when it is saturated, and this is related to the air's temperature. Warmer air requires more water vapor for it to become saturated.

When the saturation and condensation of water vapor occur on a large enough scale, they are responsible for dew, fog, clouds, and rain. Dew on the grass forms in the same way that moisture forms on a cool beverage. The air above the surface (of either the ground or the glass) is cooled to a temperature at which it becomes saturated. Once the air becomes saturated, some of the water vapor condenses onto the surface beneath the air. The temperature to which the air must be cooled is called its dewpoint.

Dewpoint is a good measure of the amount of moisture in the air because, unlike relative humidity, it is a temperature and therefore does not change with a rise or fall in air temperature. In the summer, dewpoints in Mississippi are often 60 to 70°F or more. Winter dewpoints are related to the type of air mass over the state. Cold air from Canada will have lower dewpoints than air that has been warmed and moistened by the Gulf of Mexico.

At a larger scale, whole layers of air are sometimes lifted until the temperature of the entire layer reaches its dewpoint. At that point, clouds usually form. When the condensed moisture becomes too great to remain suspended in the atmosphere, it begins to fall as precipitation.

CLOUDS AND FOG

Because of the specific ways in which different types of clouds form, they are often an indicator of what kind of weather to expect. Warm air rises at a gentle slope, and it sometimes reaches great heights before it is able to condense into clouds. That is why in winter, spring, or autumn after a few cold days, the return of warmer and typically more humid conditions will be signaled by high wispy clouds. As the warm front gets closer, the clouds usually get thicker and

RELATIVE HUMIDITY

People sometimes exaggerate when talking about weather. While discussing summertime humidity, you may have heard someone say that it was "ninety-five degrees with a hundred percent humidity!" For 100% relative humidity to be achieved, the temperature and the dewpoint (the temperature at which air becomes saturated) must be equal. In the scenario where it is 95°F with 100% relative humidity, the dewpoint would have to be 95°F also. This situation is not that common—even in the Deep South in the summer. Dewpoints along the Gulf of Mexico can often rise into the low 80s, but for most of Mississippi, even 80°F dewpoints are not common. The highest dewpoint ever recorded was 95°F in Dhahran, Saudi Arabia, as reported by Christopher Burt in the book *Extreme Weather* (2004).

Relative humidity is often close to 100% in the early morning hours, about the time when dew forms on the grass. As the day progresses, the temperature increases but the dewpoint will only change if the air mass over a region changes. So, throughout the course of the day, the relative humidity will decrease as the temperature is rising until the high temperature is reached in the late afternoon. Then relative humidity will start to increase once again until the low is reached in the early morning. Figure 4.14 shows the dewpoint, temperature, and relative humidity for a typical summer day in Mississippi in which there were no air mass changes.

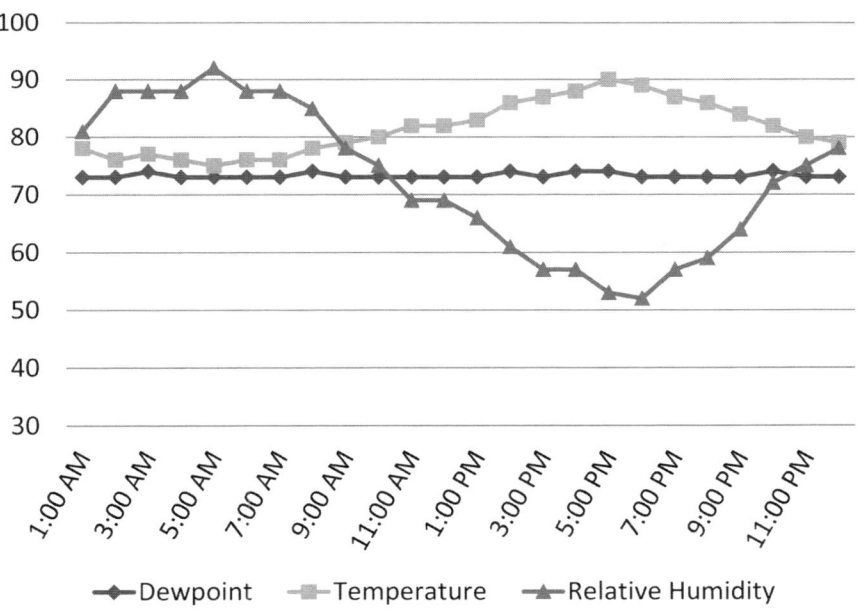

Fig. 4.14. Temperature, dewpoint, and relative humidity from a day in July. Notice that dewpoint remains in the low 70s, whereas relative humidity is highest when the temperature is the lowest at 5:00 a.m. and lowest when temperature is the highest at 6:00 p.m.

Fig. 4.15. Altocumulus clouds over the Mississippi State Campus. Altocumulus clouds are found in the middle levels of that atmosphere and often have a wavy appearance. Photo credit: K. Sherman-Morris

closer to the surface. Many people can identify the cauliflower-like appearance of the cumulonimbus cloud, which brings summer thunderstorms, as well as the gloomy gray stratus-type clouds that bring rain or drizzle. Fluffy cumulus clouds indicate instability and are fairly common. They often increase in coverage during summer as the day warms. Cumulus clouds also sometimes get thicker prior to the passage of a cold front.

Fog is another fairly common manifestation of condensation. Fog is simply a cloud at or close to the ground. For fog to form, a layer of air needs to be cooled to its dewpoint or saturated by the addition of more water vapor. This can happen in many ways, though some are more common here in Mississippi.

Fog in Ditches or Low-Lying Areas
During your travels in the early evening, you may have observed wispy sheets of fog hovering in ditches along the road or over rivers and ponds. One of the ways fog can form is if a layer of air is cooled until its temperature reaches its dewpoint. As the air starts to cool in the evening, it becomes denser. This cooler, dense air will sink into any depressions on the surface. Many rural roads are surrounded by ditches, so this becomes a prime place for the cooler air to sink. If the air cools sufficiently, fog will form.

Coastal Fog

The Gulf of Mexico is responsible for many aspects of Mississippi weather, and it also contributes to fog development. Water in the Gulf is often warmer than the adjacent land. When warmer, moist Gulf air flows over a cool land surface, fog can form. The warm air is cooled by contact with the colder ground, and if the air is cooled to its dewpoint, fog will form. Because the air originated over the Gulf of Mexico, it will be moister than air that originates over land. This causes its dewpoint to be higher, so it will not take much cooling for the dewpoint to be reached.

Pea-Soup Fog

Sometimes fog hangs on for days and seems very thick. This "pea-soup fog" is common in Mississippi when the ground is already saturated from days of rain or a recent heavy rain. The air near the surface will already be close to saturation due to the rain evaporating into the air near the ground. During the cool season, warm air occasionally settles over the cooler air at the surface. This can have two effects that help create fog. The warm air can be cooled from below as described in the previous example, and warm air in place over cooler air will act like a cap on the surface air and not allow it to rise until some other phenomenon comes along to disturb it.

Steam Fog

Steam fog is a type of evaporation fog that sometimes forms after heavy rain. The name is not entirely accurate, because for steam to be produced the air would have to reach its boiling point of 212°F. But we commonly think of any water vapor in the air that we can see as steam. Immediately after a hard rain—usually after the ground has had time to dry and warm—you may witness steam rising from the hot ground. Steam fog is often short-lived compared to the other types of fog.

DROUGHT AND FIRE: THE FAMINE

Figure 4.16 illustrates the history of wet and dry periods in Mississippi since 1895. The graph shows the monthly values of the Palmer Drought Severity Index (PDSI), which combines the elements of rainfall, temperature, and soil moisture to provide an indicator of how wet or dry a region is at any given time. The figure shows that from the beginning of the period of record (1895) through the 1940s, there were numerous, lengthy dry periods in the state. Several notable periods of drought since 1950 are also evident: 1951–1957, 1962–1967, 1980–1982, 1999–2000, and 2007.

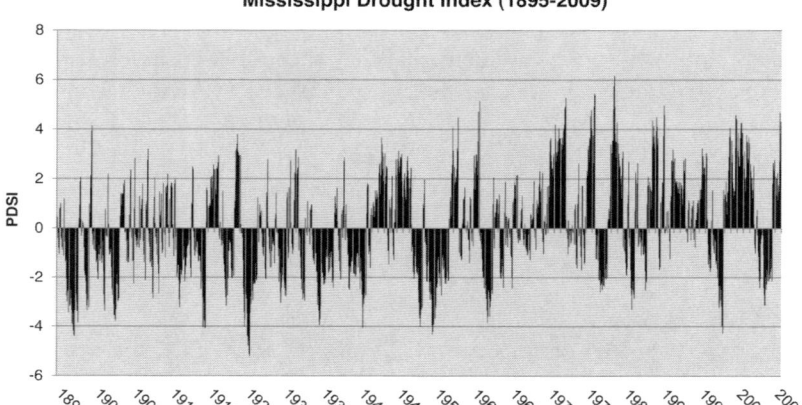

Fig. 4.16. Mississippi statewide Palmer Drought Severity Index (PDSI), 1895–2009. Severe drought occurs when the PDSI drops below –3 and extreme drought when the PDSI drops to or below –4.

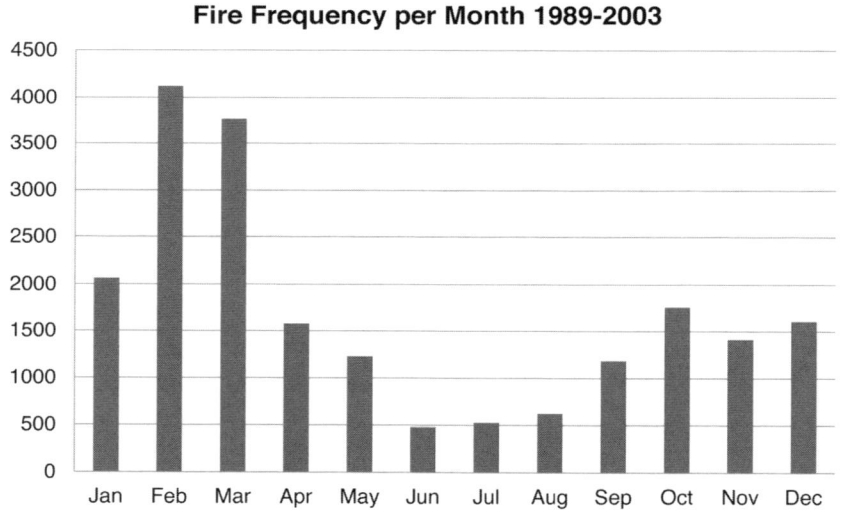

Fig. 4.17. Monthly fire frequency across southern Mississippi, 1989–2003. Image credit: John Morris

Table 4.2 lists the number of years between 1896 and 1995 that the Lower Mississippi River region had severe or extreme drought. During a severe drought, PDSI values range from –3.0 to –3.9. In an extreme drought, PDSI is less than –4.0. Some part of the region experienced a severe or extreme drought in 56 of those years, in 38 of the years drought covered more than 10% of the

Table 4.2. Number of years the Lower Mississippi River region experienced severe or extreme drought, 1896–1995

Percentage of land area in severe or extreme drought	Number of years with that percentage drought
>0%	56
>10%	38
>25%	19
>33%	15
>50%	4
>66%	1
>75%	0
>90%	0
100%	0
Data from the National Drought Mitigation Center	

region, and that the region was never as much as 75% covered by severe or extreme drought in any one year. Although droughts can occur at any time of the year in Mississippi, in general they occur during summer and are often accompanied by a heat wave. Agriculture is impacted by the summer drought and heat, particularly row crops and poultry. Pasture productivity may also be extensively reduced during drought periods, impacting livestock and having consequences in the grocery store for months after the event.

Besides causing drought and hindering the development of crops, lack of rainfall is also a precursor to fire. Mississippi has fire seasons that correspond with the transitional seasons, spring and autumn. This is the time of year when dry conditions occur at the same time as the fuel—leaf litter and dry brush, grasses and small trees. Fires are possible during the summer when a location is experiencing drought conditions; however, once the leaves fall from the trees in autumn and frost kills the undergrowth, the dead and drying vegetation provides a good source of fuel for the fires until the trees and grasses begin to grow again in the spring (NWS, 2008). Because Mississippi gets substantial rainfall in an average year (typically 50 inches or more), a rainfall deficit must exist for a heightened fire risk to exist.

The Jackson National Weather Service recently published a guide to fire weather in the Southeast that describes the conditions that lead to higher fire potential. The dry conditions may come from several months of below-normal precipitation or from a several weeks with almost no rainfall (NWS 2008). The latter was the case in the autumn of 2010. Most of the state had received little rainfall in September and October. This was coupled with warmer than normal temperatures and resulted in drought conditions. The dry conditions prompted

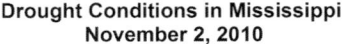

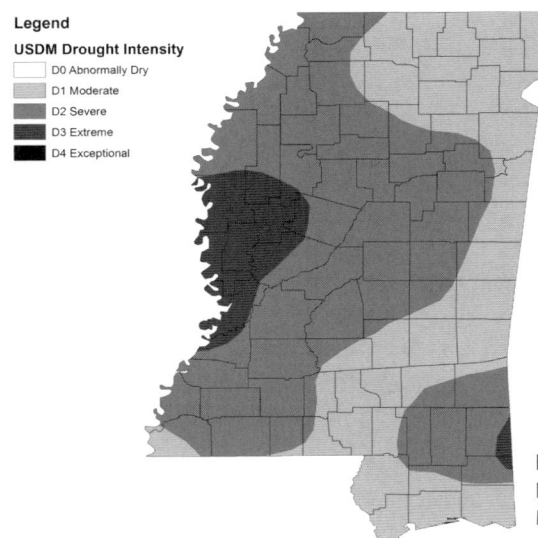

Fig. 4.18. Drought conditions for Mississippi, November 2, 2010. Data from U.S. Drought Monitor

the governor to issue a statewide burn ban in October, by which time many local communities had already issued local burn bans. Figure 4.18 is based on data from the U.S. Drought Monitor to reflect conditions on November 2, 2010.

Conditions in autumn 2010 were dry and sunny for days at a time, and such conditions are most common when the state is under surface high pressure. Typically, high pressure does not remain in place for too long a time period, especially in late September and October, when Mississippi usually starts to experience frontal systems more regularly. There are three upper air patterns that help high pressure become established across the Southeast (NWS 2008): a strong zonal flow, a northwesterly flow, or a blocking ridge. Zonal flow describes the conditions when winds in the upper atmosphere blow primarily from west to east, without much of a northerly or southerly component. This situation results in fronts that do not allow Gulf moisture to build up across the state before the passage of a cold front. Similarly, air that comes from the northwest does not tap into Gulf moisture and is also drying. The blocking ridge is associated with the Bermuda High located off the East Coast. When the ridge is centered near the Atlantic Coast, it will block low pressure systems and prevent them from moving into the region. High pressure, low humidity, and dry cold fronts that were unable to tap into the Gulf of Mexico were largely responsible for the dry conditions in 2010. The Bermuda High also blocked tropical systems from reaching the Southeast.

Further Reading

Barry, J. M. 1997. *Rising Tide: The Great Mississippi Flood of 1927 and How It Changed America*. New York: Touchstone.

Hederman, T. M. 1979. *The Great Flood*. Jackson, MS: Clarion Ledger/Jackson Daily News.

Fig. 5.1. Supercell outside of Starkville, Mississippi, April 20, 2011. Photo credit: Greg Nordstrom

5. THUNDERSTORMS, LIGHTNING STRIKES, AND TORNADOES

To Mississippians, dangerous and severe weather is a well-known part of the state's climate. The state experiences thunderstorms on an average of 81 days per year, ranging from 56 in Montgomery County to 121 in Pearl River County. These thunderstorms produce an average of almost 10,000 cloud-to-ground lightning strikes per year in each county, ranging from 5459 in Choctaw County to 19,446 in Jackson County. For the period 2001–2007, an astounding 5,643,965 cloud-to-ground lightning strikes were recorded in the state of Mississippi. In addition, an average of 27 tornadoes is reported in Mississippi annually. For the period 1950–2006, 1541 tornadoes were reported in the state, ranging from 4 in Itawamba County to 52 in Hinds County.

Mississippi's geographic location accounts for the large number of danger-ous weather events in the state. Warm, moist air masses originating in the Gulf of Mexico to the south often interact with mid-latitude cyclones and associated fronts coming into the state from the north and west. Mississippi is located in a region known as Dixie Alley, which rivals the classic Tornado Alley in the number of tornadoes per square mile. In fact, Mississippi ranks eighth ahead of both Nebraska (ninth) and Texas (tenth). Mississippi also ranks second in total tornado-related deaths, behind Texas. However, when these data are averaged over the total state population, Mississippi leads the nation in tornado-related deaths per million people. Of the top ten killer tornadoes in U.S. history, three occurred in Mississippi: in Natchez in May 1840, with 317 deaths; in Purvis in April 1908, with 143 deaths; and in Tupelo in April 1936, with 216 deaths.

Even life-long residents of the state may not realize the magnitude and spa-tial distribution of dangerous weather events in Mississippi. Advances in data availability now make it possible to quantify and visualize these aspects of the climate, enhancing awareness and safety throughout the state. Therefore, in this chapter we can provide specific information on thunderstorms, lightning strikes, and tornadoes for each county in the state.

The lightning data presented in this chapter were obtained from the National Lightning Detection Network. This dataset contains latitude and longitude co-ordinates for all cloud-to-ground lightning strikes that occurred in Mississippi between 2000 and 2007. A single thunderstorm generally produces multiple lightning strikes. The high variability in the frequency of thunderstorms and

Legend

Number of thunderstorm days

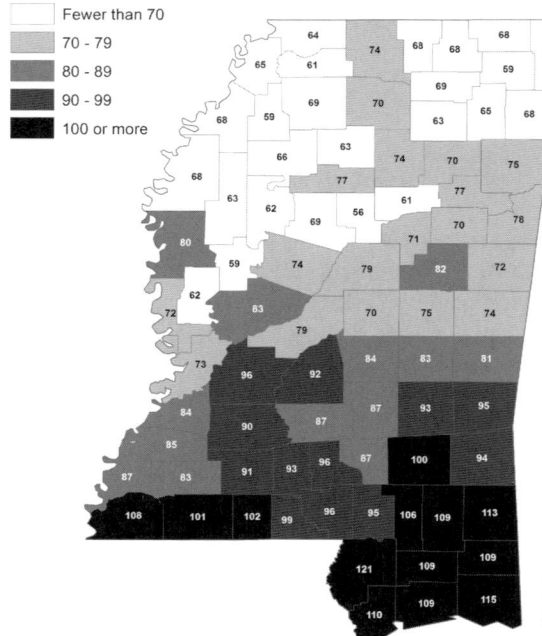

☐	Fewer than 70
▨	70 - 79
▨	80 - 89
▨	90 - 99
■	100 or more

Fig. 5.2. Average number of thunderstorm days in the state annually

the number of strikes produced could bias the calculation of the true number of thunderstorms in each county. Therefore, we instead identified the number of days on which thunderstorms occurred. A thunderstorm day is defined as a day in which at least one cloud-to-ground lightning strike was recorded.

The tornado data presented in this chapter were obtained from the Storm Prediction Center of the National Weather Service (NWS). This dataset contains all reported tornadoes, nationwide, for the period 1950–2006. The 1541 tornadoes reported in Mississippi during this 57-year period are categorized by county, month, and year. Death, injury, and economic loss data were derived from the National Climatic Data Center's Storm Event Database (NOAA 2010).

THUNDERSTORMS

A thunderstorm is any storm that produces lightning and therefore thunder. In Mississippi thunderstorms are produced by frontal activity; warm, moist summer air masses; sea breezes in the coastal region; and land-falling tropical systems that move across the state. A thunderstorm can and often does produce damaging winds, hail, heavy rainfall, and lightning. Common air mass thunderstorms lasting 60–90 minutes account for the bulk of thunderstorm activity in Mississippi. Thunderstorms that produce hail 1 inch or greater, winds 58 mph or greater, and/or a tornado are considered severe thunderstorms, which also occur in the state. The types of storms that often meet the severe criteria include the supercell storm, clusters of storms known as mesoscale convective complexes, and lines of storms known as mesoscale convective systems, which are commonly referred to as squall lines. Although the percentage of thunderstorms in Mississippi that reach severe status is not known, nationwide about 10% of thunderstorms are classified as severe and about 2–3% are classified as supercells. While severe storms garner the greatest attention, the majority of thunderstorm-related deaths are the result of flash flooding and lightning.

Table 5.1 shows the average monthly and annual number of thunderstorm days for each county in Mississippi. Two-thirds of the days in which thunderstorms occur are in the period May through September. The thunderstorms during this time period are predominantly air mass thunderstorms. Figure 5.2 depicts the spatial pattern of annual average thunderstorm days in the state.

Table 5.1. Monthly and annual county and state thunderstorm days, 2000–2008												
	Jan	Feb	Mar	Apr	May	Jun	Jul	Aug	Sep	Oct	Nov	Dec	Ann
Adams	6	4	5	6	11	11	21	4	6	6	3	4	87
Alcorn	6	1	3	8	13	6	7	9	8	1	3	3	68
Amite	1	6	8	7	10	12	24	11	9	6	2	5	101
Attala	3	4	5	5	12	4	17	13	8	3	3	2	79
Benton	6	1	5	8	10	4	9	9	8	2	3	3	68
Bolivar	2	2	2	11	11	10	6	8	6	2	4	4	68
Calhoun	3	2	1	7	13	8	12	11	7	3	3	4	74
Carroll	6	1	2	8	11	5	13	8	6	4	2	3	69
Chickasaw	3	2	1	8	11	6	13	12	7	2	2	3	70
Choctaw	1	2	3	5	9	7	17	15	6	1	3	2	71
Claiborne	6	3	5	7	9	8	18	11	9	3	2	3	84
Clarke	1	5	6	5	12	10	15	20	7	4	5	5	95
Clay	6	3	1	9	12	6	15	11	7	2	3	2	77
Coahoma	2	1	4	9	13	8	3	9	10	2	4	3	68
Copiah	6	5	5	7	12	8	19	9	8	4	3	4	90
Covington	6	5	5	5	12	9	18	11	7	3	3	3	87
De Soto	1	1	5	8	12	9	4	9	8	1	3	3	64

	Jan	Feb	Mar	Apr	May	Jun	Jul	Aug	Sep	Oct	Nov	Dec	Ann
Forrest	6	5	6	6	10	14	21	13	12	3	6	4	106
Franklin	1	5	6	6	7	7	18	12	9	5	3	4	83
George	6	6	5	5	9	14	24	14	12	7	3	4	109
Greene	6	7	5	6	10	13	27	16	13	2	4	4	113
Grenada	1	2	2	7	11	6	11	9	8	14	2	4	77
Hancock	6	7	9	6	5	14	25	15	10	6	4	3	110
Harrison	6	6	5	6	8	16	24	17	10	7	2	2	109
Hinds	6	3	5	8	13	9	18	14	9	5	3	3	96
Holmes	1	5	3	9	10	6	12	11	7	3	3	4	74
Humphreys	6	1	3	7	8	4	7	8	7	2	3	3	59
Issaquena	6	2	4	7	9	7	13	7	5	4	4	4	72
Itawamba	1	2	1	9	10	8	12	13	5	2	3	2	68
Jackson	6	7	5	6	9	16	23	20	11	6	3	3	115
Jasper	2	5	6	6	10	8	19	18	7	4	3	5	93
Jefferson	6	3	7	8	10	9	16	11	5	2	4	4	85
Jeff. Davis	6	5	5	6	10	11	20	14	8	4	4	3	96
Jones	6	5	5	5	12	12	20	16	8	3	4	4	100
Kemper	1	4	3	5	12	5	15	16	6	2	1	4	74
Lafayette	1	1	2	8	13	9	11	7	9	3	2	4	70
Lamar	6	6	6	5	9	10	19	12	10	5	4	3	95
Lauderdale	1	4	5	6	12	7	16	14	7	3	3	3	81
Lawrence	6	5	5	6	10	10	22	12	8	4	2	3	93
Leake	1	3	3	7	11	6	13	11	6	4	3	2	70
Lee	2	1	1	10	11	6	10	10	7	1	2	4	65
Leflore	1	2	3	10	10	4	10	7	7	2	3	3	62
Lincoln	6	5	5	7	10	7	20	12	9	3	3	4	91
Lowndes	1	2	2	7	12	6	13	10	6	14	2	3	78
Madison	1	2	5	7	11	7	16	14	6	3	4	3	79
Marion	6	6	6	6	11	10	18	13	10	4	3	3	96
Marshall	2	1	5	8	13	6	11	8	10	2	4	4	74
Monroe	3	1	1	9	12	8	14	12	6	2	3	4	75
Montgomery	1	1	1	6	6	5	13	9	7	1	3	3	56
Neshoba	6	5	3	6	13	4	14	12	5	3	2	2	75

	Jan	Feb	Mar	Apr	May	Jun	Jul	Aug	Sep	Oct	Nov	Dec	Ann
Newton	1	5	4	7	12	8	13	15	7	4	4	3	83
Noxubee	1	3	5	5	12	7	15	13	4	1	3	3	72
Oktibbeha	1	2	2	4	10	6	17	14	6	3	3	2	70
Panola	2	1	4	8	13	9	7	7	10	1	3	4	69
Pearl River	6	6	6	5	9	14	27	19	14	8	4	3	121
Perry	6	6	6	5	11	13	23	14	12	2	6	5	109
Pike	6	6	8	6	10	11	23	12	11	4	2	3	102
Pontotoc	2	1	1	7	12	8	8	9	8	1	2	4	63
Prentiss	2	1	3	8	6	10	10	6	6	1	3	3	59
Quitman	1	1	3	9	11	8	4	6	6	3	3	4	59
Rankin	6	4	5	9	10	6	17	17	7	5	3	3	92
Scott	1	5	5	8	12	5	16	14	8	5	3	2	84
Sharkey	6	2	4	5	8	5	9	7	7	2	3	4	62
Simpson	6	5	5	6	11	9	16	13	6	4	2	4	87
Smith	1	5	5	7	11	7	17	16	8	4	2	4	87
Stone	6	6	5	5	10	17	22	14	10	6	4	4	109
Sunflower	1	1	2	11	11	6	7	8	6	2	4	4	63
Tallahatchie	1	2	3	11	12	5	6	9	8	3	3	3	66
Tate	1	1	4	8	12	9	5	5	9	1	3	3	61
Tippah	6	1	3	7	13	5	10	8	8	2	2	3	68
Tishomingo	1	1	3	8	12	9	9	10	6	1	3	4	67
Tunica	6	1	3	9	12	8	3	7	9	2	3	2	65
Union	6	1	2	7	13	6	10	10	7	1	2	4	69
Walthall	6	7	7	7	8	10	20	12	11	4	3	4	99
Warren	1	3	6	7	9	7	12	11	7	4	3	3	73
Washington	6	3	4	11	12	9	9	11	6	2	3	4	80
Wayne	6	5	5	4	12	11	17	13	9	3	5	4	94
Webster	1	2	1	5	12	4	12	10	8	2	2	2	61
Wilkinson	1	4	6	7	12	13	26	15	10	6	4	4	108
Winston	1	3	5	8	12	6	20	14	6	2	3	2	82
Yalobusha	2	1	1	8	12	6	10	7	7	3	3	3	63
Yazoo	6	4	3	7	10	5	18	12	8	3	3	4	83
Statewide	4	3	4	7	11	8	15	12	8	3	3	3	81

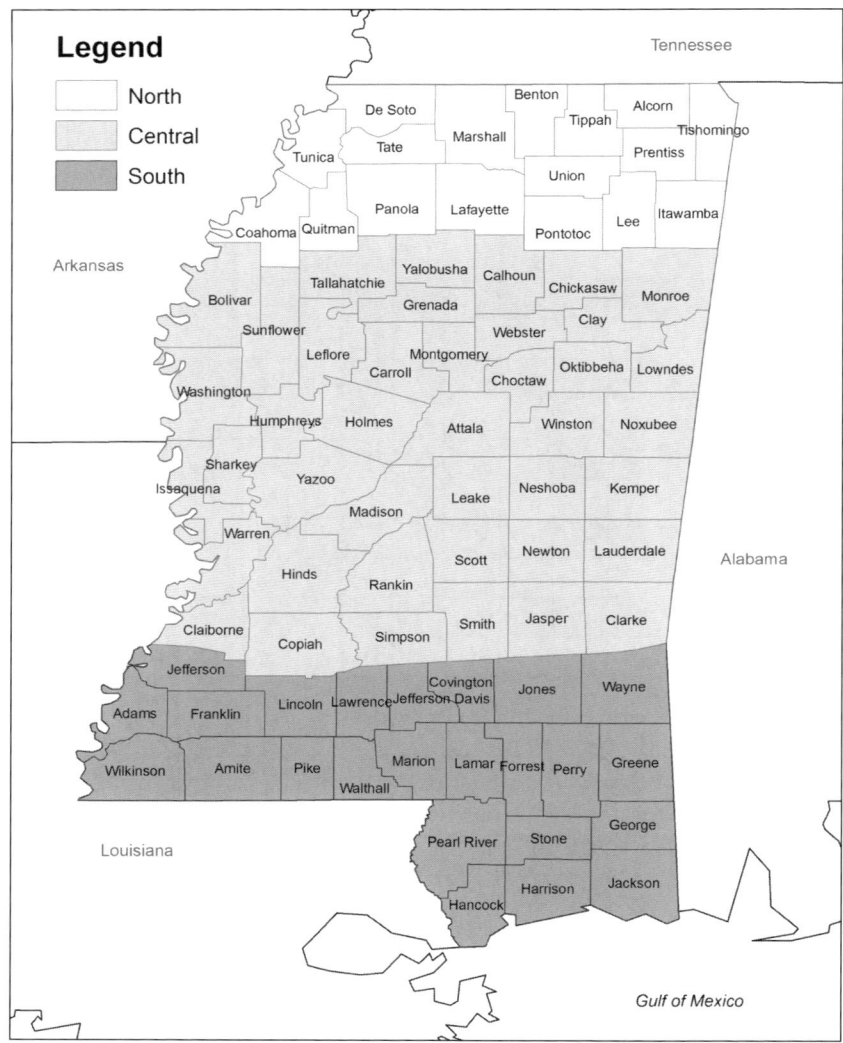

Fig. 5.3. Counties in the northern, central, and southern regions

In Figure 5.3, counties are grouped together to form southern, central, and northern regions in the state. The southern region is comprised of 24 counties south of a line spanning from Jefferson to Wayne Counties. The northern region is comprised of 17 counties north of a line from Coahoma to Itawamba Counties. The central region is comprised of the 41 counties between the southern and northern regions. Among these regions, there is a decreasing south-to-north gradient in the average annual thunderstorm days, with 100.8 in the

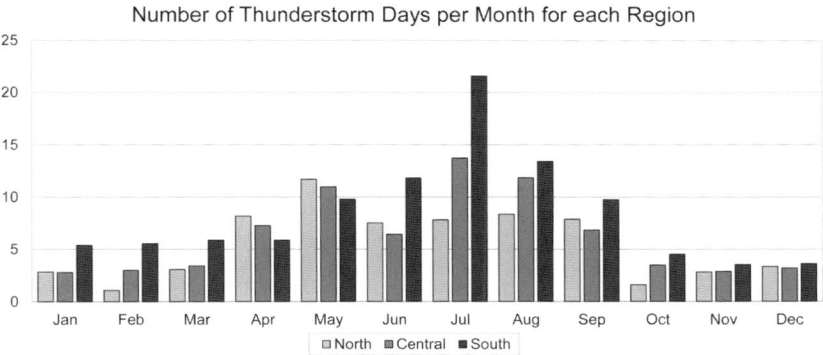

Fig. 5.4. Monthly distribution of thunderstorm days for the northern, central, and southern regions of Mississippi

southern region, 75.9 in the central region, and 66.2 in the northern region. Figure 5.4 represents the distribution of thunderstorm days by month for the three regions. This figure illustrates the high number of thunderstorm days during the summer in all regions; three-fourths of the thunderstorm days occur between April and September. It also shows the predominance of thunderstorm days in the southern region in all but 2 months of the year.

The gradient in number of thunderstorm days from the coast to the northern part of the state is a response to increasing distance from the Gulf of Mexico. Maritime tropical air masses that originate in the Gulf are the source of moisture and warm air, necessary ingredients for thunderstorm formation. Seasonal differences in the processes that lift the warm, moist air to initiate thunderstorm development are evident in Figure 5.4. Thunderstorm days in the northern region peak in April and May and outnumber those in the southern region, most likely due to frontal activity. By contrast, in the southern region thunderstorm days peak in the middle of summer, most likely due to the inherent instability of the air mass and sea breeze processes.

LIGHTNING

The electrical discharge of a thunderstorm is lightning, which can travel from the base of a cloud at nearly 130,000 mph and reach temperatures three to four times that of the surface of the sun. Lightning is the most deadly feature of a thunderstorm, and the extent to which it occurs in Mississippi is much more widespread and frequent than generally recognized. For example, in an average year 50 lightning strikes occur within a 1-mile radius of the flagpole at the center of the Mississippi State University campus. Within 3 miles, that number exceeds 1200. By comparison to any other weather phenomenon, these numbers

Fig. 5.5. Lightning strike along the northern Gulf Coast. Photo credit: Grady Dixon

are extraordinarily large, highlighting the threat to life and property posed by lightning across the state.

Table 5.2 lists the number of cloud-to-ground lightning strikes that occurred during 2001–2007 in each county in Mississippi, monthly and annually. The total in the state during this period was a shocking 5,643,965, with an annual average of 806,281 strikes. However, from 1994 to 2010 there were only 203 instances where lightning was reported to have caused deaths, injuries, or damage (National Oceanic and Atmospheric Administration Storm Data). During this period, lighting was responsible for at least 15 deaths, 35 injuries, $10.4 million in property damage, and $9000 in crop damage. The majority of these reported events were structures that were struck by lightning and burned. Based on the number of lightning strikes that occur, it is likely that lightning events are underreported in the state and are also underestimated as a weather threat to Mississippians.

Whereas Table 5.1 lists only the number of thunderstorm days, the data in Table 5.2 are the actual number of strikes recorded by the National Lightning Detection Network. One thunderstorm is capable of producing multiple strikes, so the number of lightning strikes is much larger than the number of thunderstorms. The seasonality of lightning in Mississippi is evident in Figure 5.6, which shows that nearly half the year's total occurs in July and August and

Table 5.2. Monthly and annual county and state cloud-to-ground lightning strikes, 2001–2007

	Jan	Feb	Mar	Apr	May	Jun	Jul	Aug	Sep	Oct	Nov	Dec	Annual Avg.	Total
Adams	269	2342	3089	3133	7765	9729	12500	12070	2423	1337	4113	1620	8629	60400
Alcorn	588	719	1570	2953	4650	3710	12645	9165	2327	2309	1426	339	6057	42401
Amite	446	3772	7387	7439	12496	19112	31295	19852	4098	1732	1804	2269	15957	111702
Attala	682	2067	2539	9904	8051	8952	20530	14348	1983	1653	5303	1330	11049	77342
Benton	198	502	1499	4997	5959	4485	11378	7442	2430	2294	1827	846	6265	43857
Bolivar	1418	2240	4408	7927	23891	12989	18408	14444	2528	2530	4432	1911	13875	97126
Calhoun	380	1441	4467	5189	14401	13944	13922	12804	2044	1999	2467	1098	10594	74156
Carroll	404	1040	2561	4540	7911	8888	15767	11374	1512	1499	2443	1095	8433	59034
Chickasaw	452	674	3158	4125	10884	7622	11251	13653	1539	1146	2114	937	8222	57555
Choctaw	501	1251	1599	3207	5441	3600	10252	6697	1240	499	3346	578	5459	38211
Claiborne	392	1241	3019	3110	12919	7925	15291	10129	2275	682	3961	1366	8901	62310
Clarke	623	2190	5282	7353	8737	10992	20013	22249	2963	1521	1733	1609	12181	85265
Clay	432	759	2534	3072	7565	6684	11299	8775	1185	1016	3154	419	6699	46894
Coahoma	625	886	4321	5145	17174	9975	14522	9517	1397	2057	2874	872	9909	69365
Copiah	507	4338	4381	4595	13438	17401	27829	20358	5459	1429	3836	1631	15029	105202
Covington	173	1108	4223	4472	4404	6556	13994	9955	2128	327	1824	553	7102	49717

	Jan	Feb	Mar	Apr	May	Jun	Jul	Aug	Sep	Oct	Nov	Dec	Annual Avg.	Total
De Soto	376	638	2987	4465	9146	5175	10634	8074	1544	1625	3539	533	6962	48736
Forrest	419	677	3320	5576	4785	10021	18057	14257	2762	961	1626	1062	9075	63523
Franklin	524	3480	4593	3311	10031	14670	21399	14903	3938	1145	2463	1403	11694	81860
George	353	707	2761	5103	5644	12756	20678	18501	3541	1724	1625	1045	10634	74438
Greene	764	1214	4201	6961	8164	15068	27579	18692	5240	1691	4333	1558	13638	95465
Grenada	259	1271	3074	3754	10734	7398	11043	9605	1193	1355	2489	709	7555	52884
Hancock	368	1245	3220	5558	7112	12454	22390	15154	4412	2031	1721	1193	10980	76858
Harrison	457	1712	2905	8249	8491	19452	28874	23401	6728	2726	2343	1226	15223	106564
Hinds	509	1702	4522	8129	15177	12923	22861	15949	3050	1308	5661	1778	13367	93569
Holmes	523	1676	2311	8659	8680	12526	14679	11613	1317	1952	3218	1232	9769	68386
Humphreys	286	994	1489	4022	4864	7184	8953	5729	1748	994	1290	912	5495	38465
Issaquena	294	1010	1385	4855	5986	5208	8763	6230	1770	770	1341	1043	5522	38655
Itawamba	770	660	2393	5709	11320	6184	13620	9676	2652	2048	2634	807	8353	58473
Jackson	263	1490	4560	10431	8101	24647	37109	35687	7525	2629	2254	1427	19446	136123
Jasper	594	1989	4531	6200	8536	8067	23448	18358	3580	968	2685	820	11397	79776
Jefferson	508	2366	3539	3700	12417	10979	17772	13496	3915	1063	4363	1578	10814	75696
Jefferson Davis	180	1753	3828	3607	3948	7195	15645	9617	2408	300	1518	557	7222	50556

	Jan	Feb	Mar	Apr	May	Jun	Jul	Aug	Sep	Oct	Nov	Dec	Annual Avg.	Total
Jones	634	1468	7310	8280	10515	13449	23068	16082	2013	761	3551	1491	12660	88622
Kemper	466	1887	3136	9139	15031	7591	18995	18232	2242	3446	1628	1026	11831	82819
Lafayette	765	1629	3966	9837	15675	9314	20689	14850	2951	3205	3065	930	12411	86876
Lamar	391	709	3369	4617	4795	9441	15371	12357	4177	838	2142	973	8454	59180
Lauderdale	506	2094	4520	8372	14761	10049	21592	16372	2394	2557	1932	911	12294	86060
Lawrence	420	2164	3504	3460	5676	8296	13323	11153	1621	508	1399	645	7453	52169
Leake	411	1657	1650	9820	5143	6692	16765	13868	1756	853	2340	777	8819	61732
Lee	414	783	3108	5663	9687	5055	13007	10914	2354	2045	1546	677	7893	55253
Leflore	524	1454	2870	6093	11489	9129	12986	8806	2286	1572	3423	971	8800	61603
Lincoln	643	3319	5122	4278	7434	14882	19543	17394	3119	829	1693	1357	11373	79613
Lowndes	452	1360	2181	4867	5568	6538	15313	11846	1953	1206	4800	410	8071	56494
Madison	408	1704	2372	7556	8946	10146	17865	14610	2587	919	4611	1126	10407	72850
Marion	243	1364	5898	5688	4702	8713	14648	13965	2380	991	2075	921	8798	61588
Marshall	606	632	2914	9124	12199	7087	18415	12246	3632	3152	3751	1306	10723	75064
Monroe	962	1301	4068	8452	14306	10041	20683	12312	4453	1693	4703	495	11924	83469
Montgomery	311	675	1877	2317	4489	5445	11339	8995	823	1006	1795	856	5704	39928
Neshoba	477	1894	2372	11134	7349	5682	14317	16723	1493	1765	1698	935	9406	65839

	Jan	Feb	Mar	Apr	May	Jun	Jul	Aug	Sep	Oct	Nov	Dec	Annual Avg.	Total
Newton	636	2047	3319	7458	9539	6501	18705	14161	2195	1293	1229	987	9724	68070
Noxubee	438	1986	2640	7146	6102	7206	19696	16386	1907	1485	2725	1123	9834	68840
Oktibbeha	607	1201	2203	3713	4120	5391	14007	10777	948	604	4250	343	6881	48164
Panola	1005	1345	3396	9282	16720	11674	18565	12343	2730	3129	4168	1149	12215	85506
Pearl River	1591	1383	5020	11595	9921	17827	29967	28231	4531	2680	5021	2038	17715	119805
Perry	527	1158	4019	7278	6857	14150	22044	20281	3682	1671	2841	1464	12282	85972
Pike	382	1558	3787	4270	5913	7770	13671	10430	2239	711	1851	1122	7672	53704
Pontotoc	469	1350	3844	6530	12110	6272	15556	13059	2578	2264	2007	578	9517	66617
Prentiss	499	517	1757	4381	6238	3990	12841	8270	1188	1430	1165	519	6114	42795
Quitman	431	583	3067	4579	10741	5837	10806	7270	859	1751	2671	578	7025	49173
Rankin	727	3089	3424	7810	12846	14298	28387	22451	3481	1182	3519	1788	14715	103002
Scott	877	1980	3450	7976	9206	7727	20076	19528	2366	548	2316	1213	11038	77263
Sharkey	370	978	1565	3955	7011	3991	9845	6162	1404	570	1424	1031	5472	38306
Simpson	708	2972	3647	3630	6135	10376	19671	16693	2276	697	2452	1018	9968	69775
Smith	603	2497	3840	5507	8507	8135	23163	17198	2778	1019	2412	933	10942	76592
Stone	488	705	2870	6676	3786	12187	19407	18231	3233	1569	2227	883	10323	72262
Sunflower	610	1400	3388	8251	12945	10828	16429	11527	2813	2264	3854	1170	10783	75479

	Jan	Feb	Mar	Apr	May	Jun	Jul	Aug	Sep	Oct	Nov	Dec	Annual Avg.	Total
Tallahatchie	739	1713	3544	7853	19879	12697	16366	15555	2751	2212	2899	818	12432	87026
Tate	389	266	1882	5218	9393	4561	7833	8323	1411	1274	2414	707	6239	43671
Tippah	409	765	1405	4546	6706	4963	16962	8477	1459	2272	1719	970	7236	50653
Tishomingo	535	730	2036	3232	5478	3769	13428	8876	2251	1473	1017	326	6164	43151
Tunica	161	610	2346	4309	9141	4654	8987	7498	1670	1804	2252	558	6284	43990
Union	509	747	2210	6541	9010	4938	11755	11751	1932	2954	1513	697	7794	54557
Walthall	246	1361	4654	5261	3709	7770	12958	10357	1635	855	1743	850	7343	51399
Warren	435	1409	2520	6648	12207	8818	15742	9103	2428	1058	3072	1482	9275	64922
Washington	464	1498	3584	7875	7897	8148	14365	7830	2492	1281	3233	900	8510	59567
Wayne	658	1766	6659	9184	10227	14810	25595	24272	3578	772	2944	1323	14541	101788
Webster	458	539	2601	2284	6085	8705	11963	7946	845	1063	1788	606	6412	44883
Wilkinson	448	4115	6310	5889	11175	17501	26153	16642	3202	2858	3572	2495	14337	100360
Winston	472	1909	2082	6105	6691	5087	18789	16228	1357	1151	3909	1049	9261	64829
Yalobusha	412	1172	3982	5197	12829	9418	14095	10295	2173	1496	2644	563	9182	64276
Yazoo	826	1803	3150	9227	12054	8488	19875	12104	2599	1657	4221	1841	11121	77845
Statewide	42299	124400	276094	497553	757765	770508	1420021	1110254	210079	125713	222984	86295	806281	5643965

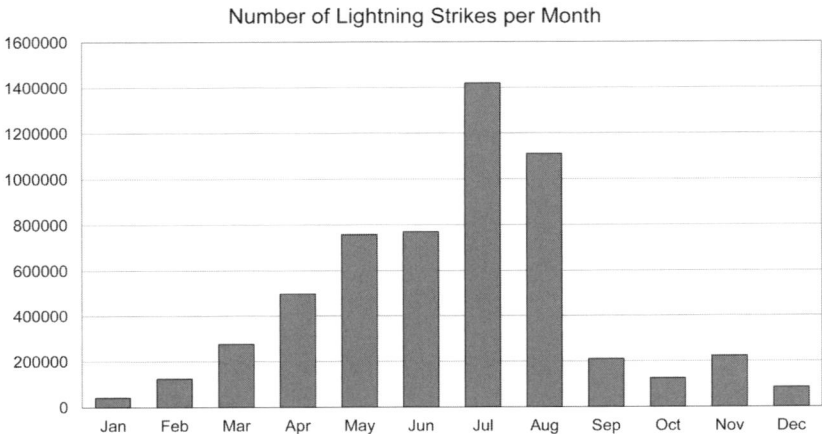

Fig. 5.6. Monthly lightning strikes statewide, 2001–2007

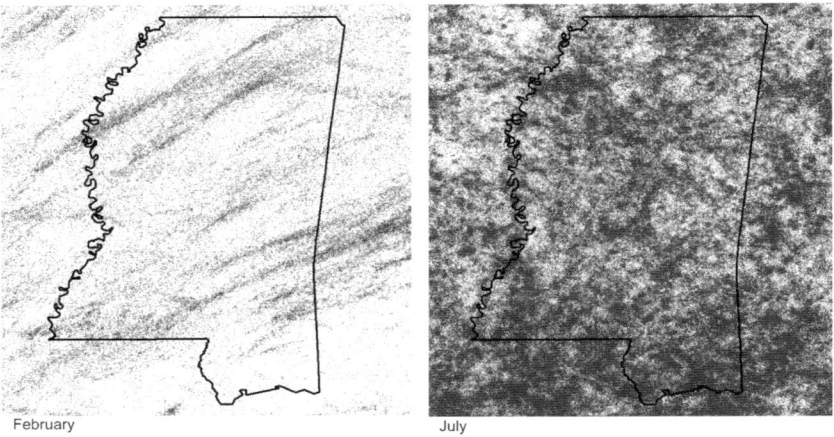

Fig. 5.7. Seasonal differences in the spatial distribution of lightning strikes in Mississippi. The linear patterns in the February map are the result of thunderstorms associated with cold fronts, whereas air mass thunderstorms during the summer are spread throughout the state.

almost three-fourths of the year's total occurs between May and August. By contrast, December and January together have less than 1% of the year's total. Figure 5.7 illustrates this distinct seasonality and explains the difference in the origin of the lightning. The spatial distribution of February lightning strikes is related to the movement of individual thunderstorms with the passage of cold fronts, producing linear patterns of strikes. The distribution of lightning strikes in July, however, is representative of the ever-present afternoon thunderstorms in the warm, muggy summer air masses.

SKYWARN

SKYWARN is a volunteer program established by the National Weather Service (NWS) to provide critical information during severe weather. There are almost 300,000 trained SKYWARN spotters in the United States. The main duty of SKYWARN volunteers is to identify and describe severe weather they see occurring in their local area. If you would like to get involved with your local SKYWARN, you can contact the Warning Coordination Meteorologist at your local NWS office (http://www.stormready.noaa.gov/contact.htm). Training is free.

Spotters learn useful information including severe weather vocabulary, how to spot a tornado in the field, how to observe rotation in clouds, and how to estimate hail size using standard measurements. The verification of a tornado's location on the ground can help the NWS protect life and property through timely and descriptive warnings.

Unlike the tornado data presented below, the lightning and thunderstorm day data do not rely upon human observation of the event. Instead, the locations of the lightning strikes are measured by the automated National Lightning Detection Network. As a result, there are no inherent population biases such as more reports near towns and cities and fewer in rural areas.

TORNADOES

A tornado is a violently rotating column of air that is in contact with the ground and the base of a cumulonimbus cloud. Although tornadoes can exhibit many shapes, they are most commonly recognized by a funnel shape. On the small scale, tornadoes are the most destructive of all atmospheric phenomena. Extremely strong pressure gradients associated with the rotating structure of these storms produce wind speeds that can exceed 200 mph and cause nearly total destruction of the area hit by the storm.

Mississippi is located in the middle of Dixie Alley, a region that experiences a proportionate number of tornadoes as occur in the better known Tornado Alley in the Great Plains. Given the nature of the forested landscape in Mississippi as compared to the relatively flat and open landscape of the Great Plains, the hazard to life and property from tornadoes in the state is considerable. Averaged over total area, Mississippi has the same number of tornadoes each year as Illinois, Texas, and Nebraska (Grazulis 1993). Only Oklahoma, Kansas, and Iowa have more than Mississippi. When considering only strong and violent tornadoes (EF4 and EF5 ratings), Mississippi is behind only Oklahoma. Furthermore, mobile homes comprise nearly 20% of total housing units in Mississippi, second

Fig. 5.8. All tornado occurrences in Mississippi, 1950–2010.
Image credit: Grady Dixon

in the nation only to South Carolina (Brooks and Doswell 2001). These mobile homes are subject to greater damage at lower wind speeds when compared to traditionally built structures, as indicated by the Enhanced Fujita Scale (Texas Tech University 2006).

Reporting of tornadoes relies upon human observation, and therefore the tornado database exhibits a population bias. The degree of bias is related to the population density of a region, the terrain (i.e., observation distance), the existence or absence of organized storm-spotting organizations, and the road network of the region (Kelly et al. 1985). Consequently, rural areas may be underrepresented in the data of spatial and temporal distributions of tornadoes in Mississippi presented here.

Figure 5.8 shows the pattern of tornado touchdowns and paths in the state during 1950–2010. A tornado path has clear beginning and ending locations with continuous or nearly continuous damage between the two, whereas a tornado touchdown is a single point location of isolated damage. Tornado events mapped in Figure 5.8 include both touchdowns in a single county and paths that crossed multiple counties. The figure shows the characteristic southwest-to-northeast direction of movement evident in the paths. The map also shows that tornadic activity appears to be greater in the southern and central regions of the state.

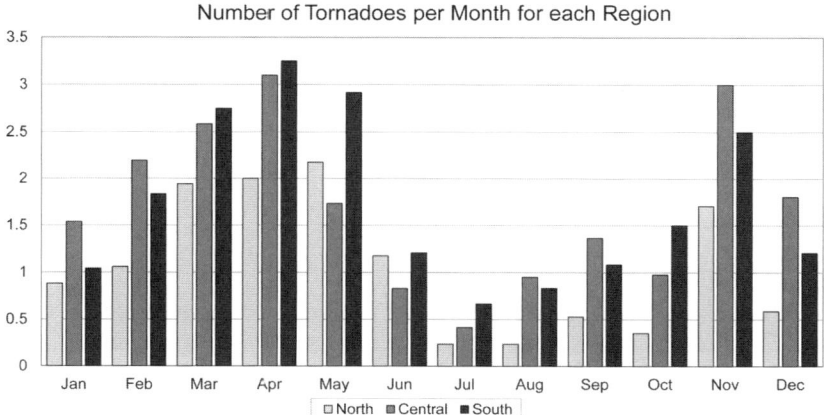

Fig. 5.9. Total tornado occurrences in Mississippi by month, 1950–2006

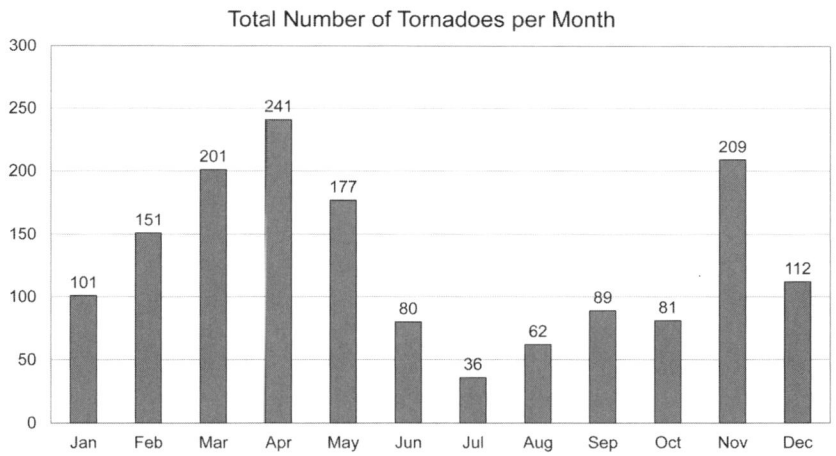

Fig. 5.10. Annual average of tornado occurrences in the northern, central, and southern regions of the state

The distribution of tornadoes by month is shown in Figure 5.9. Unlike the summer maximum of thunderstorm and lightning activity in the state, tornadoes exhibit peaks in both early spring and late autumn. These two periods represent the times when frontal passages are most frequent in the state. Figure 5.10 displays the distribution of tornadoes by regions in the state, showing the predominance of tornadoes in the south and central regions and once more emphasizing the extent to which the Gulf of Mexico influences severe weather in Mississippi.

Table 5.3. Monthly and annual county and state totals of tornado events, 1950–2006												
	Jan	Feb	Mar	Apr	May	Jun	Jul	Aug	Sep	Oct	Nov	Dec	Annual
Adams	1	0	3	2	2	2	0	0	0	0	1	0	11
Alcorn	1	2	1	3	5	1	0	0	0	0	2	0	15
Amite	0	0	4	1	4	0	1	0	0	1	3	3	17
Attala	2	1	5	0	2	3	1	1	2	1	3	1	22
Benton	0	1	2	0	3	0	0	0	1	0	0	0	7
Bolivar	5	2	5	4	4	2	1	0	7	1	4	1	36
Calhoun	2	2	3	5	0	1	0	0	0	0	0	1	14
Carroll	2	2	0	4	1	0	0	0	1	0	1	1	12
Chickasaw	0	1	1	2	2	2	1	0	1	0	2	3	15
Choctaw	1	1	0	2	0	0	0	0	1	0	0	0	5
Claiborne	2	2	4	3	0	0	0	1	3	1	4	1	21
Clarke	2	0	2	1	6	0	0	4	1	1	5	0	22
Clay	1	0	3	2	1	1	0	1	0	1	0	0	10
Coahoma	0	0	1	7	4	3	1	0	0	1	2	1	20
Copiah	1	3	3	2	0	0	1	3	0	3	3	5	24
Covington	0	0	2	9	1	0	1	0	1	0	2	1	17
De Soto	2	0	3	4	1	2	0	1	0	0	4	0	17
Forrest	1	2	5	2	6	1	1	1	0	0	0	2	21
Franklin	1	0	0	1	0	1	0	0	2	2	2	0	9
George	1	3	2	1	2	1	0	0	1	2	0	0	13
Greene	0	3	3	2	1	0	0	0	0	0	2	2	13
Grenada	0	5	1	1	1	1	0	0	0	0	2	4	15
Hancock	0	2	1	6	9	2	5	2	2	2	3	1	35
Harrison	0	3	6	6	10	3	1	5	9	3	3	0	49
Hinds	5	7	6	7	3	1	2	2	6	1	9	3	52
Holmes	2	2	2	4	2	0	0	1	0	2	2	2	19
Humphreys	2	1	1	3	1	0	0	1	5	0	2	3	19
Issaquena	1	1	2	3	0	0	0	0	0	0	2	1	10
Itawamba	0	0	1	2	1	0	0	0	0	0	0	0	4
Jackson	3	5	2	4	9	4	3	3	2	0	2	2	39
Jasper	2	3	3	1	1	0	0	0	1	1	0	1	13
Jefferson	1	1	3	2	2	0	0	0	4	1	1	0	15

	Jan	Feb	Mar	Apr	May	Jun	Jul	Aug	Sep	Oct	Nov	Dec	Annual
Jefferson Davis	2	1	1	0	3	0	0	0	0	2	2	3	14
Jones	2	3	9	6	2	4	0	0	0	5	6	7	44
Kemper	0	0	1	1	1	0	1	2	0	0	4	3	13
Lafayette	0	2	4	1	1	1	0	1	0	0	1	0	11
Lamar	2	2	4	1	0	0	0	2	1	3	5	0	20
Lauderdale	2	3	6	1	3	1	1	6	2	1	3	3	32
Lawrence	2	3	0	5	0	0	0	3	1	0	3	0	17
Leake	2	3	4	0	2	0	0	2	1	2	3	1	20
Lee	1	0	3	4	4	0	0	0	1	2	1	1	17
Leflore	4	2	1	8	4	1	1	0	3	0	2	4	30
Lincoln	2	2	3	10	1	0	0	1	0	5	3	2	29
Lowndes	1	3	5	4	1	1	0	0	1	3	2	2	23
Madison	1	5	8	5	6	0	0	0	1	0	7	3	36
Marion	1	3	3	2	1	2	1	1	0	2	3	1	20
Marshall	2	1	3	4	3	1	0	0	2	1	0	0	17
Monroe	3	3	3	3	2	1	0	0	1	0	1	1	18
Montgomery	2	1	1	0	1	0	0	0	0	3	4	0	12
Neshoba	1	3	4	3	1	1	0	4	0	2	5	1	25
Newton	1	4	2	2	1	2	1	3	1	0	5	2	24
Noxubee	0	0	3	0	1	1	0	0	1	0	7	0	13
Oktibbeha	2	1	1	0	1	0	1	0	2	0	1	4	13
Panola	1	1	2	1	1	1	1	0	0	0	1	1	10
Pearl River	2	4	4	3	8	0	3	2	1	3	4	0	34
Perry	0	3	2	2	1	2	0	0	0	0	1	2	13
Pike	2	1	2	6	0	2	0	0	0	0	3	1	17
Pontotoc	2	2	2	2	4	0	0	0	0	0	2	1	15
Prentiss	2	0	1	1	1	1	0	0	0	1	5	3	15
Quitman	0	0	0	0	1	1	0	0	0	0	3	0	5
Rankin	2	10	2	11	3	0	0	0	2	2	6	3	41
Scott	0	3	0	3	4	3	0	1	0	2	2	2	20
Sharkey	1	0	4	2	0	1	0	1	2	1	3	1	16
Simpson	1	2	5	7	2	0	2	1	1	2	5	2	30
Smith	3	2	3	9	1	2	0	1	2	4	5	2	34

	Jan	Feb	Mar	Apr	May	Jun	Jul	Aug	Sep	Oct	Nov	Dec	Annual
Stone	0	1	0	4	1	2	0	0	1	2	3	1	15
Sunflower	1	1	1	5	2	0	2	0	3	0	4	0	19
Tallahatchie	0	0	3	4	1	2	0	0	0	0	0	3	13
Tate	1	0	0	2	2	0	1	1	0	0	0	1	8
Tippah	0	1	0	2	3	3	0	0	2	0	0	0	11
Tishomingo	1	0	1	1	1	2	0	0	0	0	5	0	11
Tunica	0	5	0	1	1	1	0	0	0	0	0	1	9
Union	0	2	5	1	0	0	0	0	1	0	0	0	9
Walthall	1	1	4	3	3	1	0	0	1	0	3	0	17
Warren	3	2	0	3	2	2	1	1	2	2	5	5	28
Washington	2	4	0	7	0	3	0	0	2	1	3	1	23
Wayne	0	1	3	0	3	1	0	0	0	1	2	0	11
Webster	0	1	3	0	1	1	0	0	1	0	2	0	9
Wilkinson	1	0	0	0	1	1	0	0	0	2	3	1	9
Winston	0	1	0	2	1	0	0	0	0	0	3	1	8
Yalobusha	1	1	1	0	2	0	0	0	0	0	0	1	6
Yazoo	0	2	4	3	4	1	1	3	0	3	2	2	25
Statewide	101	151	201	241	177	80	36	62	89	81	209	112	1541

Monthly and annual tornado occurrences for each county are presented in Table 5.3. Between 1950 and 2006, only two counties, Hinds and Lauderdale, experienced tornadoes in every month of the year. On the other extreme, Itawamba County experienced tornadoes only in March, April, and May. Figure 5.11 shows the spatial distribution of all tornado events by county. This figure highlights the areas of apparent greatest tornado frequency as the Delta and the central and coastal regions.

April is the month when the greatest number of tornadoes occurs in the state. Figure 5.12 shows the spatial distribution of tornadic activity for that month in Mississippi, once more indicating the pattern of greatest occurrence in the Delta and central and coastal regions. The spatial distribution of tornadic activity in November, the secondary peak of activity for the year in Mississippi, is shown in Figure 5.13. At this time of year, the greatest concentration of tornadoes occurs in the central part of the state from Jackson to south of Highway 82. The month with the minimum amount of tornadic activity in Mississippi is July. Figure 5.14 shows that the spatial distribution in July is the same as the more active months, but fewer events occur across the state.

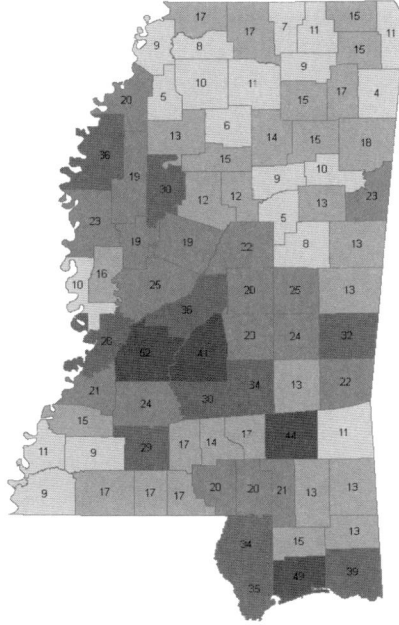

Fig. 5.11. All tornadoes reported in Mississippi, 1950–2006. Counties are labeled with the number of tornadoes reported in each.

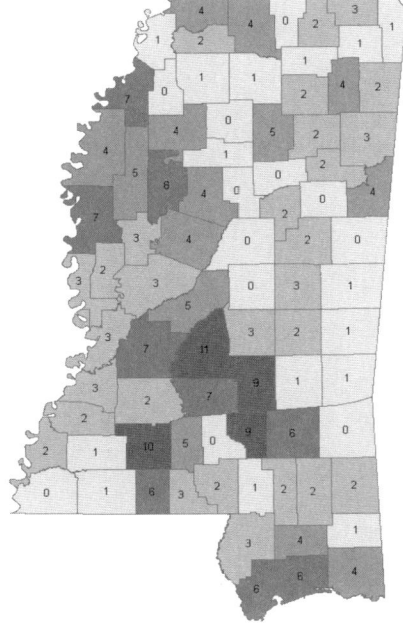

Fig. 5.12. April tornadoes reported in Mississippi, 1950–2006. Counties are labeled with the number of tornadoes reported in each.

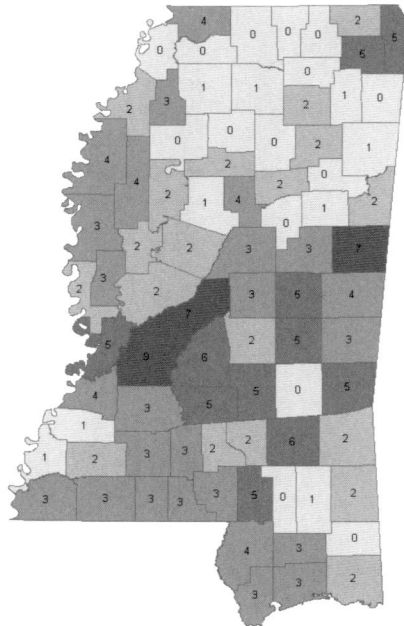

Fig. 5.13. November tornadoes reported in Mississippi, 1950–2006. Counties are labeled with the number of tornadoes reported in each.

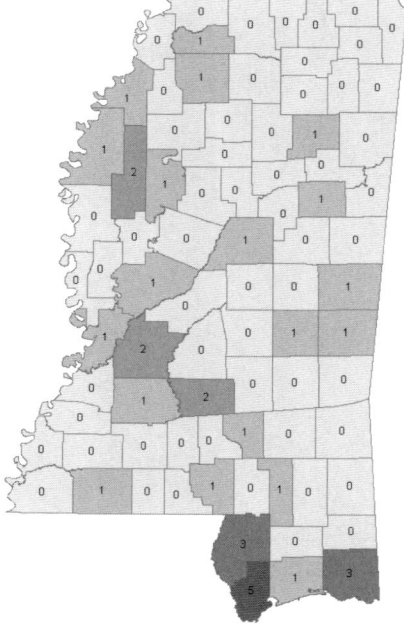

Fig. 5.14. July tornadoes reported in Mississippi, 1950–2006. Counties are labeled with the number of tornadoes reported in each.

HISTORICAL TORNADOES

May 7, 1840: Natchez Tornado

One of the first and deadliest tornadoes documented in Mississippi was the Natchez tornado of May 7, 1840. This storm killed 317 people in Natchez and the surrounding area, and it is the second deadliest tornado in U.S. history. Only the Tri-State tornado of 1924 killed more people. This mile-wide tornado entered the Mississippi River approximately 7 miles southwest of the city of Natchez, Mississippi. The tornado stayed in or near the river channel until striking the northern portion of the city. One description of the tornado stated "the air was black with whirling eddies of walls, roofs, chimneys, and huge timbers from distant ruins . . . all shot through the air as if thrown from a mighty catapult" (Grazulis 1993). Forty-eight people were killed in Natchez, and an additional 269 were killed in the Mississippi River when flatboats and steamers were destroyed. Many speculate that the death toll in Natchez was grossly underestimated, as the death of slaves was rarely accounted for. Given the rural nature of the landscape and the comparatively early developmental stage and population count of the city of Natchez at that time, the amount of economic damage has never been estimated. However, based on the conservatively reported death count, the storm is considered to have been a rare EF5 tornado. A storm of that intensity striking that area now would cause much more economic devastation.

April 23–24, 1908: Purvis Tornado, the Dixie Tornado Outbreak

According to the Birmingham NWS, at least 34 tornadoes touched down from Texas to Georgia during the Dixie tornado outbreak. One of the areas worst-affected by the tornadoes was Purvis, Mississippi. This tornado crossed from Louisiana into Mississippi, clearing a path 2 miles wide according to the *Encyclopedia of Disasters* (Gunn 2008). When the tornado hit Purvis, the town was leveled, and only 7 of 150 houses remained standing.

April 5, 1936: Tupelo Tornado

This tornado, an estimated F5, is still considered one of the deadliest in U.S. history. The large tornado moved east-northeast across the northern portion of the city of Tupelo, and it was responsible for completely destroying more than 200 homes. Entire families were killed; in one case, all 13 members of a family were while inside a single home. According to estimates, 233 people were killed in the city of Tupelo by this single tornado. However, because records were never kept on the number of injuries suffered by the African American community (newspapers of the time only published injuries affecting White residents), some suspect the death toll was considerably higher. One-hundred fifty railroad box cars were brought to Tupelo to be used a temporary housing. The city's movie theater served as a hospital, and the popcorn machine was

Fig. 5.15. Gloster Street in Tupelo, Mississippi, April 1936. Photographer unknown

used to sterilize medical instruments. Elvis Presley survived this tornado; he was a year old at the time.

March 3, 1966: The Candlestick Park Tornado

This tornado developed in Hinds County and moved northeast, initially through mostly rural areas before entering the Jackson City limits, where it destroyed the Candlestick Park Shopping Center. According to the Jackson NWS website, eyewitnesses reported cars being thrown more than a half mile. The tornado was even more destructive in Scott County. It continued to cause damage and deaths even after it left the state, crossing into Alabama. Fifty-seven people were killed by this tornado in Mississippi, about half in Scott County alone. According to the NWS, the Candlestick Park tornado was the deadliest, most damaging, and longest track tornado in the 20th century in central Mississippi.

February 21, 1971: Mississippi Delta Outbreak

This outbreak produced 19 tornadoes, including three with tracks covering multiple counties in Mississippi. According to a map created by the Tornado History Project, one long-path F4 tornado touched down in Issaquena County and exited the state along the Tennessee border, west of Walnut in Tippah County. An F5 also cut through the Delta. The death toll for this outbreak in Mississippi stands at 110 lives lost and almost 1500 injured.

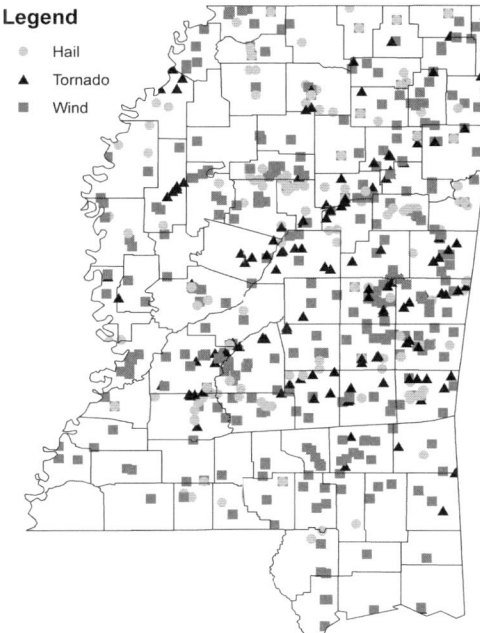

Legend
- ● Hail
- ▲ Tornado
- ■ Wind

Fig. 5.16. Storm reports made during April 2011. Data are from the Storm Prediction Center, National Oceanic and Atmospheric Administration, Department of Commerce

April 23–24, 2010: Yazoo City Tornado

The Yazoo City storm of April 23–24, 2010, provides perspective on fatalities and economic destruction brought to the state by severe weather. The series of storms that formed during that outbreak produced seven tornadoes. The most significant of these began in Madison Parish, Louisiana, and tracked northeastward into Oktibbeha County, Mississippi. At its greatest intensity the tornado was rated EF4, with winds estimated at 170 mph in Yazoo and Holmes Counties. The tornado track was an astonishing 149 continuous miles on the ground, and it reached a width of 1.75 miles. The tornado was actually on the ground for nearly 3 hours, killing 10 and injuring 146 Mississippians, and damaging or destroying hundreds of structures. While a path of 149 miles is an extreme occurrence, it should be noted that Mississippi leads the nation in tornado path length.

April 2011: Severe Weather Events

The month of April 2011 was prolific in terms of the number of severe weather events. Figure 5.16 shows a preliminary count of the number of storms (tornado, wind, or hail) reported to the Storm Prediction Center for Mississippi. Many of these were tornado reports. There were five severe weather events spanning 10 days that brought severe storms to every corner of the state. April

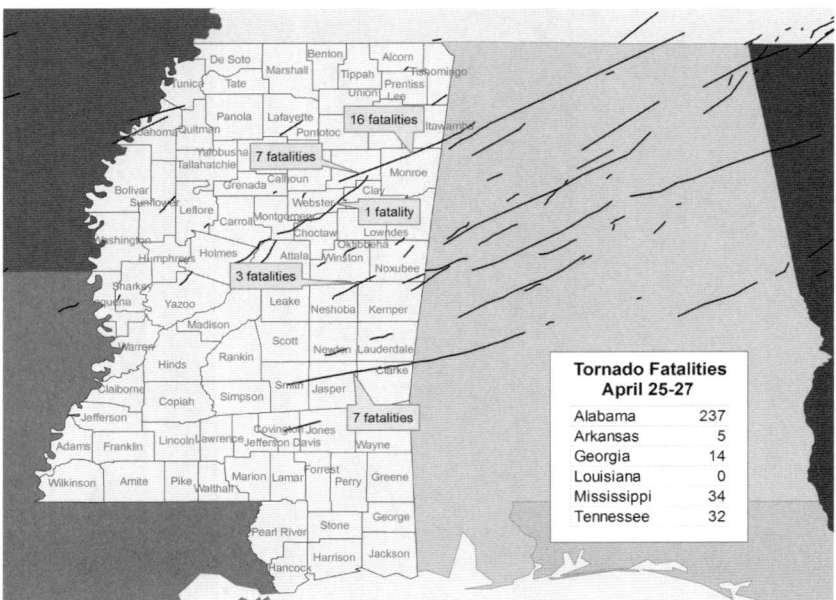

Fig. 5.17. Estimated paths of the individual tornadoes that occurred April 24–29, 2011. Most fatalities are based on National Oceanic and Atmospheric Administration /National Weather Service estimates and may not be final.

4 was largely a wind event, with 135 wind reports. An example of the type of report in the storm reports database is this one from Cleveland: "steeple blown off a church on Bishop Road (JAN)." (The three letters at the end indicate the NWS office for the report, in this case, Jackson.) April 11 also saw a wind event, with 22 reports. Many of the reports made on April 15 were tornadic, and included 19 reported injuries.

April 25–27 was the worst episode of severe weather that month. During this three-day period, there were 114 tornado reports. This does not mean that there were 114 tornadoes, because tornadoes are reported multiple times along their path. However, the day did set records. Including all tornadoes (not just those forming in Mississippi), this outbreak produced the most tornadoes to occur in a 24-hour period. Two EF5 tornadoes, the highest rating on the EF scale, formed within an hour of each other, both causing fatalities and destruction. (The last time an EF5 tornado touched down in Mississippi was the Candlestick Park tornado of May 3, 1966.) Figure 5.17 shows the preliminary tracks of the tornadoes that occurred on April 27 with the number of fatalities caused by each. The first EF5 formed in Neshoba County north of Philadelphia, causing extensive damage along its path and three fatalities once it moved into Kemper County. Another tornado that afternoon caused seven fatalities as it crossed Smith, Jasper,

Fig. 5.18. An EF3 tornado outside of Scooba, Mississippi, April 15, 2011. Photo Credit: Tim Wallace

and Clarke Counties. The most deadly tornado, another EF5, occurred in Smith-ville, Mississippi, in Monroe County. The Smithville tornado touched down at approximately 3:45 p.m. Large swaths of the town were devastated by the tornado. Mayor Gregg Kennedy described the scene at a press conference, "Our town is flat. . . . To see what happened in 10 seconds—it was gone" (Johnson et al. 2011). Among the buildings destroyed by the tornado were the town hall, post office, police station, Smithville Baptist Church, Smithville United Methodist Church, and at least 18 homes (LeCoz 2011; Memphis NWS 2011).

CONCLUSIONS

Severe weather has profound impacts on life and property in Mississippi each year. During the period 2000–2007, there was an average annual number of lightning strikes in the state of more than 800,000, and lightning killed one citizen and injured two more. The lightning strikes also caused over $650,000 of damage each year. During the period 1950–2006, the annual average number of tornadoes was 27. Each year, an average of six people were killed and another 90 were injured by tornadoes. Tornadoes also caused over $15.2 million in eco-nomic damage yearly. For perspective, the lightning damage averaged about $0.82 of economic loss per strike, whereas each tornado produced an average of about $560,000 of economic loss.

However, comparing the tornado damage from the same eight-year period as the lightning (2000–2007), there was an average of 57 tornadoes, two deaths, 45 injuries, and $35 million in economic losses each year. The increased average number of tornadoes during this period may be a result of better detection and

TORNADO SAFETY

If there is a tornado watch issued for your county, this means that conditions are favorable for storms to produce tornadoes sometime within the several-hour period the watch is issued. Watches are issued by the Storm Prediction Center in Norman, Oklahoma. During the watch, forecasters at one of the local NWS forecast offices (Jackson, Memphis, Mobile, or New Orleans) will be watching the radar for the counties in their warning areas. If the radar indicates a location where a tornado might be about to occur, or if a member of SKYWARN or someone in an official position relays a report of a tornado or funnel cloud, the NWS will issue a tornado warning. The warning is much more immediate than the watch and means you should take cover immediately if you are in the warning area.

The safest place you can go during a tornado is a basement. Most houses in Mississippi do not have basements, so the next best thing would be an interior room on the lowest floor of your house. The idea is to place as many walls between you and the tornado as possible. If you live in a mobile home, the best thing to do would be to take shelter at another location, such as the house of a friend or relative with a site-built home. If there is no place you can walk to once the warning is issued, then the safest thing to do would be to find a sturdier structure to go to before the warning is even issued. Some places have a community tornado shelter. Some companies also sell storm shelters that you can bury in the ground. This is another good option.

One thing you do not want to do is open your windows "to equalize the pressure." This action was based on outdated information about how tornadoes work. It is not true, and you could be seriously injured if you take the time to do this before a tornado strikes.

reporting rather than an actual increase in tornado occurrence. The reduction in death and injuries may be attributed to advances in tornado detection and communication technology leading to better warnings, as well as to increased media coverage and public awareness. The increase in economic loss during the more recent eight-year period may be the direct result of increased population, urban growth, and related infrastructure in the state.

Because Mississippians are exposed to so much severe weather, safety precautions and awareness should continue to be promoted across the state. In the case of thunderstorms and lightning, the 30/30 rule is endorsed by the NWS. This rule states that when lightning is observed, you should begin to count seconds until the thunder is heard. If the number of seconds is less than 30, you should go indoors and stay indoors for 30 minutes after the last thunder is heard. If tornadoes are possible in an area, the NWS will issue a tornado watch for that area. This means that conditions for tornado formation are favorable and present and people should begin to monitor the situation by tuning to local media or a National Oceanic and Atmospheric Administration weather radio.

If a tornado is indicated by radar or spotted by an observer, a tornado warning will be issued. At this time, immediate action is necessary. Citizens should have a plan already developed that includes moving to the interior of the lowest floor of their residence and remaining there until the storm has passed. Understanding the threat of severe weather in Mississippi and the vulnerability to those threats can help to reduce the death, injury, and economic loss experienced in the state each year.

Further Reading

Edwards, R. Storm Prediction Center. The Online Tornado FAQ. http://www.spc.noaa.gov/faq/tornado/
The Tornado Project Online. http://www.tornadoproject.com/

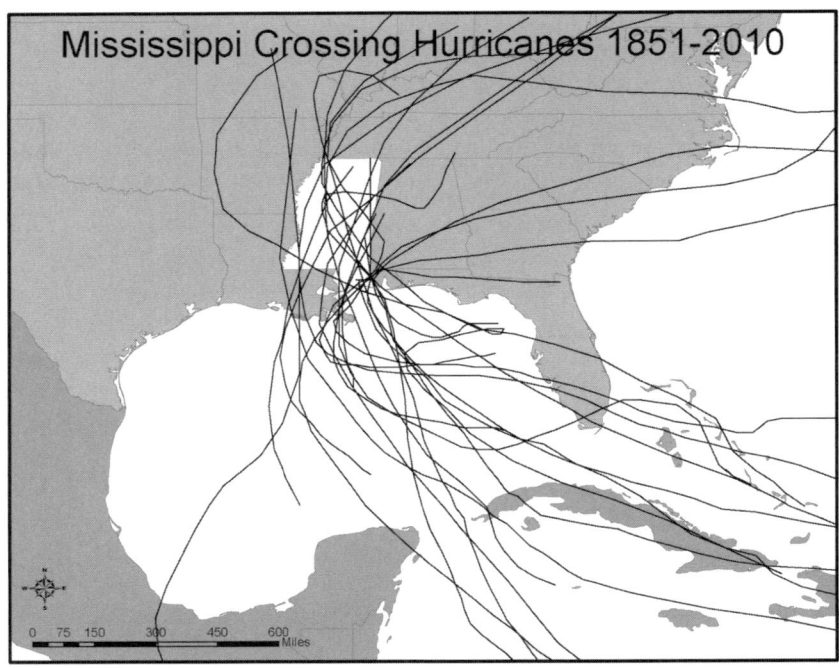

Fig. 6.1. Paths of all storms that attained hurricane strength and at some point crossed the state of Mississippi, 1851–2010. Based on National Oceanic and Atmospheric Administration best-track data set

6.HURRICANES

The Gulf, like a provoked and angry giant, can awake from its seeming lethargy, overstep its conventional boundaries, invade our land, and spread chaos and disaster. During this hurricane season, we turn to You, O loving Father. Spare us from past tragedies whose memories are still so vivid and whose wounds seem to refuse to heal with the passing of time.
—PRAYER FOR HURRICANE SEASON (ARCHDIOCESE OF NEW ORLEANS)

Another weather hazard in Mississippi is the tropical cyclone. While hurricanes are primarily a coastal hazard, they have crossed the state with hurricane force as far north as Meridian and Greenville after moving through Alabama or Louisiana. Figure 6.1 shows the paths of those tropical cyclones that at one time were hurricanes that crossed the state. Not all of these storms made landfall in one of Mississippi's three coastal counties, and they were not all hurricanes when they entered the state.

To be considered a hurricane, a tropical cyclone must have sustained wind speeds of at least 74 mph. This makes it a category 1 hurricane on the Saffir-Simpson Hurricane Wind Scale. This scale was developed in 1969 by meteorologist and then director of the National Hurricane Center, Bob Simpson, and engineer Herb Saffir. In its first four decades, the Saffir-Simpson Hurricane Scale also related wind speed with storm surge height and the hurricane's central pressure. In 2009, winds were set apart from the other factors and the scale was renamed the Saffir-Simpson Hurricane Wind Scale. Central pressure is useful for estimating wind speed, and the intensity of a hurricane is related to the height of its storm surge. However, many other factors, such as the shape of the coastline, the forward speed and direction of the hurricane, and the depth of the waters offshore all influence a hurricane's possible storm surge. So there was not always a perfect relationship between wind speed and storm surge. To make the scale less confusing to the public, the current scale only provides a measure for the strength of the hurricane's winds. Table 6.1 shows the range of wind speed and likely damage associated with each category of hurricane.

Table 6.1. Saffir-Simpson Hurricane Wind Scale		
Category	Wind Speed	Possible Effects
1	74–95 mph	Damage (to people, animals, or property) from flying debris, branches down, and damage to signs, fences, or older mobile homes. Minor damage to roofing material, siding, windows, and doors. Damage to power lines and poles.
2	96–110 mph	Considerable damage to mobile homes and vegetation. Major roof or siding damage possible on frame houses. Toppled trees may block roads. Lengthy power outages may be expected.
3	111–130 mph	Poorly constructed homes destroyed and isolated structural damage to residences. A high risk of injury or death to people or animals from debris. Older metal buildings or unreinforced masonry buildings may fail.
4	131–155 mph	Widespread destruction of older and even newer mobile homes. Roof failure on well-built homes. Top floors of apartment buildings will experience structural damage and most windows of high rise buildings will be blown out. Power outages may last for weeks or months.
5	over 155 mph	Complete roof failure on all types of buildings. A large percentage of frame houses, apartment buildings, and industrial buildings will be destroyed. The area may be uninhabitable for weeks or months.

HURRICANE SEASON

Mississippians who live along the coast watch the weather forecast from June 1 through November 30 with a bit more interest, perhaps, than those living in other parts of the state. These 6 months mark the Atlantic hurricane season. Hurricanes can occur outside of hurricane season, but these are the months when the vast majority of hurricanes occur. Figure 6.2 shows the seasonal variation in frequency of tropical cyclones in the North Atlantic. From 1850 through 2009, hurricanes actually only made landfall in Mississippi from July through October (see Figure 6.3), and most landfalls occurred in August or September.

The Gulf of Mexico provides enough water warm for hurricanes to develop throughout most of this season. In fact, the Gulf is the location with the best conditions for hurricanes at the beginning of the season in June. As temperatures warm up throughout the whole tropical Atlantic Ocean, hurricane formation shifts eastward. At the peak of hurricane season from mid-August to mid-September, hurricanes commonly originate all the way from the Gulf of Mexico to the western coast of Africa. By November, however, temperatures in the Gulf and atmospheric conditions are no longer as favorable for hurricanes to form in the Gulf of Mexico, so the preferred location for their origin shifts south to the Caribbean Sea off the coasts of Central America and the West Indies. The

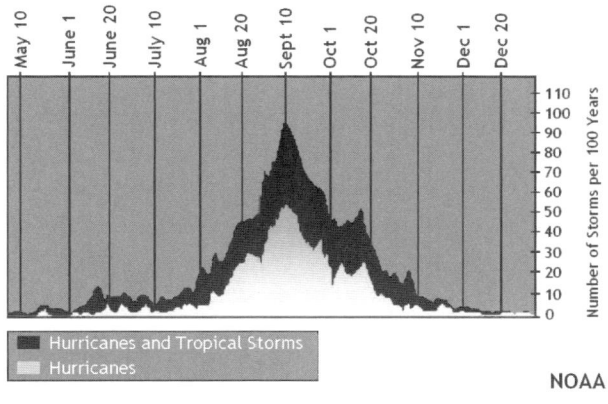

Fig. 6.2. The seasonal variation in tropical cyclone frequency shows the peak in mid-September. Image credit: National Oceanic and Atmospheric Administration, Department of Commerce

NOAA

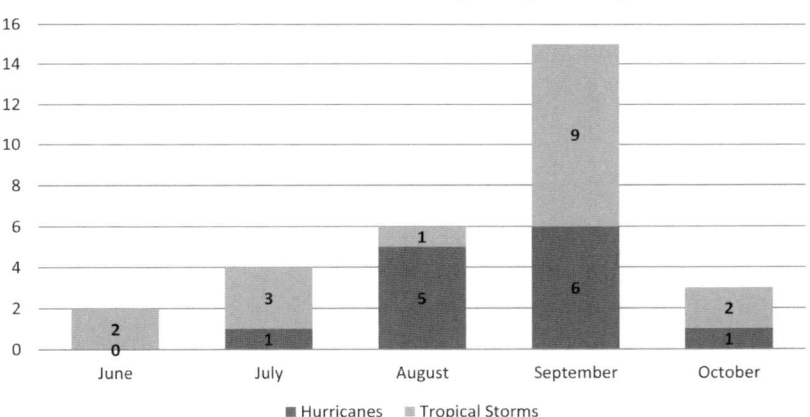

Fig. 6.3. Seasonal variation in Mississippi tropical cyclones, 1851–2010. Based on National Oceanic and Atmospheric Administration best-track data set

end of hurricane season in Mississippi corresponds with the beginning of the autumn cold front season. This is the time of year after the first significantly colder air mass invades the state.

In Figure 6.4, National Oceanic and Atmospheric Administration (NOAA) maps illustrate how the favored locations for hurricanes changes throughout the season. No tropical cyclone has made landfall in Mississippi in November, and the November map shows the only likely track is in the Caribbean, moving away from the United States.

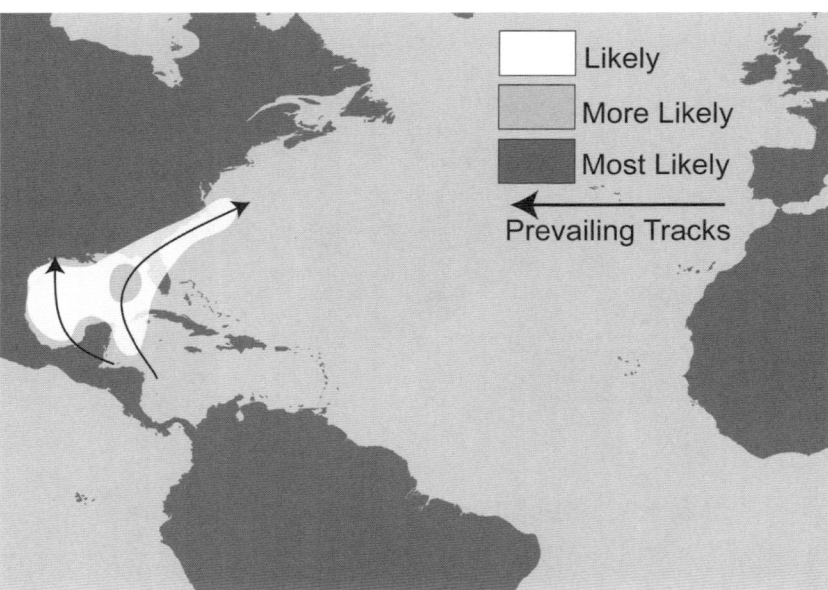

Fig. 6.4a. Most likely hurricane tracks in June. Recreated based on National Oceanic and Atmospheric Administration/National Hurricane Center image

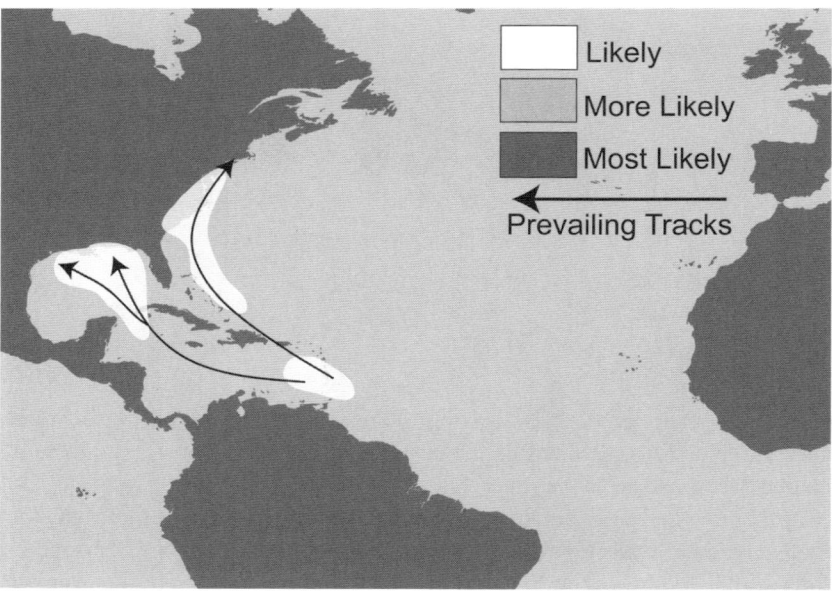

Fig. 6.4b. Most likely hurricane tracks in July. Recreated based on National Oceanic and Atmospheric Administration/National Hurricane Center image

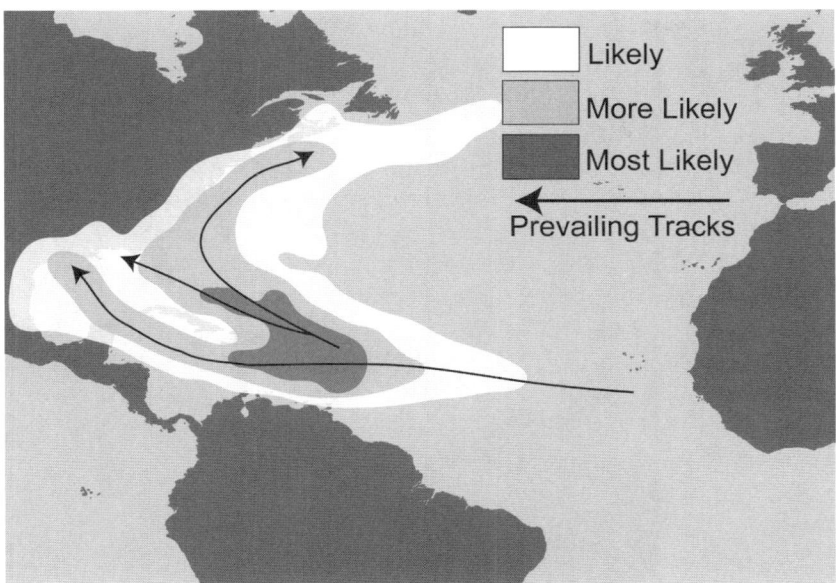

Fig. 6.4c. Most likely hurricane tracks in August. Recreated based on National Oceanic and Atmospheric Administration/ National Hurricane Center image

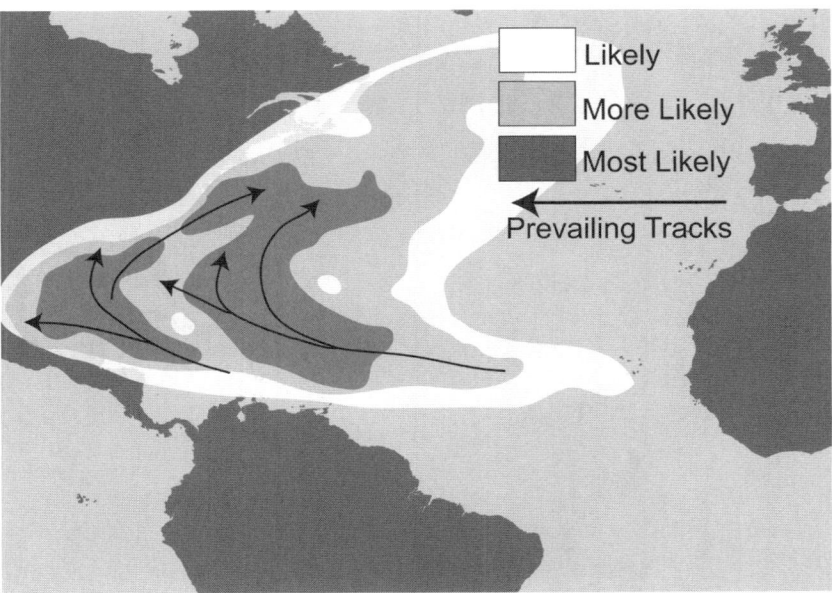

Fig. 6.4d. Most likely hurricane tracks in September. Recreated based on National Oceanic and Atmospheric Administration/ National Hurricane Center image

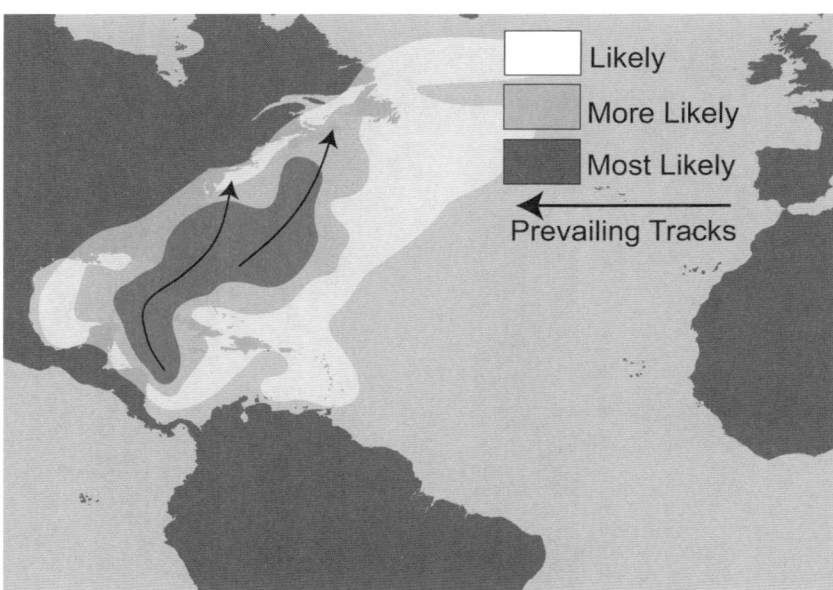

Fig. 6.4e. Most likely hurricane tracks in October. Recreated based on National Oceanic and Atmospheric Administration/ National Hurricane Center image

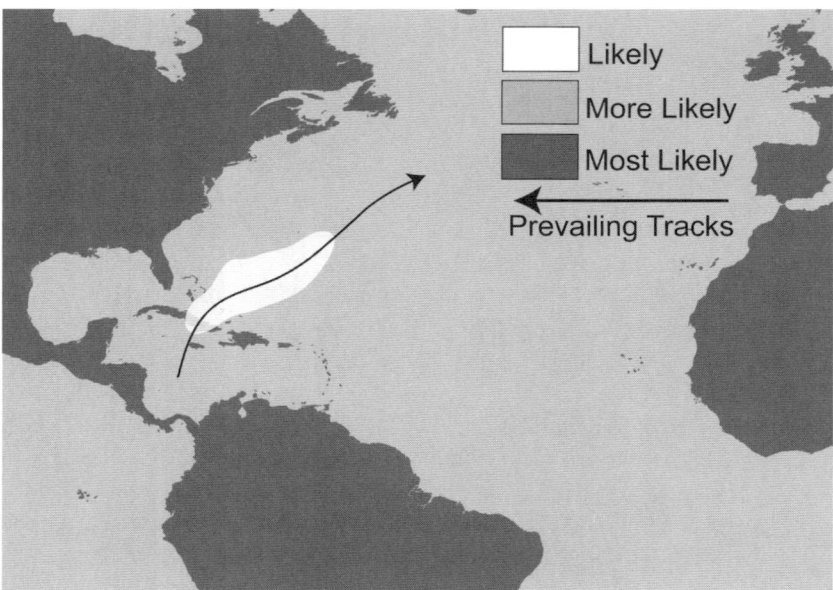

Fig. 6.4f. Most likely hurricane tracks in November. Recreated based on National Oceanic and Atmospheric Administration/ National Hurricane Center image

HURRICANE INGREDIENTS

Hurricanes occur during a particular season, when the ingredients for a hurricane to form and grow are the most ideal. First, hurricanes need a large, deep supply of warm water. Water temperature is important because hurricanes get their energy from the cycle of water evaporating from the ocean surface and then condensing into clouds in the atmosphere. Evaporation is enhanced when water temperatures are warm. The depth of warm water is important for the sustenance or growth of a hurricane because as the hurricane travels over water, it mixes the water beneath it, which can bring cooler deep water to the surface. Water temperature must be at least 80°F for hurricanes to develop. Northern Gulf water temperatures range, on average, from the low 80s in June to the low 70s in October. Water temperature is at its highest in July and August, when it is in the mid-80s, on average. The Gulf Loop Current, which is a branch of the Gulf Stream, can bring even warmer water to the Gulf of Mexico. At one point in its cycle, the current only skirts the southern Gulf of Mexico. At another point, it extends far north toward Mississippi. Sometimes a pocket of very warm water transported by the current gets separated from the main flow and becomes stranded in the Gulf for several weeks. This can cause a hurricane to intensify if it passes over the warm pocket, as happened in Hurricane Katrina.

Also important is a lack of strong wind shear. Wind shear is caused by winds that are moving at different speeds or directions at different levels of the atmosphere. To determine whether vertical wind shear is too great for a hurricane to develop, meteorologists look at the atmospheric winds at heights between 1 and 7.5 miles (which correspond to the 850 and 200 millibar pressure levels). Winds moving more than approximately 22 mph differently from the bottom to the top of this vertical area tend to inhibit hurricane development. One phenomenon that can cause the winds to be sheared is the jet stream. In October, the cold front that passes through the state and brings noticeably cooler air is associated with a large trough in the atmosphere that corresponds with the location of the jet stream. If the fast-moving jet stream is located over the Gulf or off the East Coast, it can inhibit the development of hurricanes. Once a hurricane is well developed, a greater amount of shear is necessary to break apart the storm. Other factors that can cause greater wind shear include the El Niño pattern and another strong air current off the western coast of Africa. Wind shear is bad for hurricane formation and development because it forces upper portions of the hurricane (or would-be hurricane) away from its heat source—the warm water—and does not allow the storm to vent aloft.

The Coriolis force, which is the result of the Earth's rotation and causes air to be deflected toward the right in the Northern Hemisphere, is required for hurricanes to form. Mississippi is far enough from the equator that this force is always sufficient. This factor only prevents hurricane formation between 5°

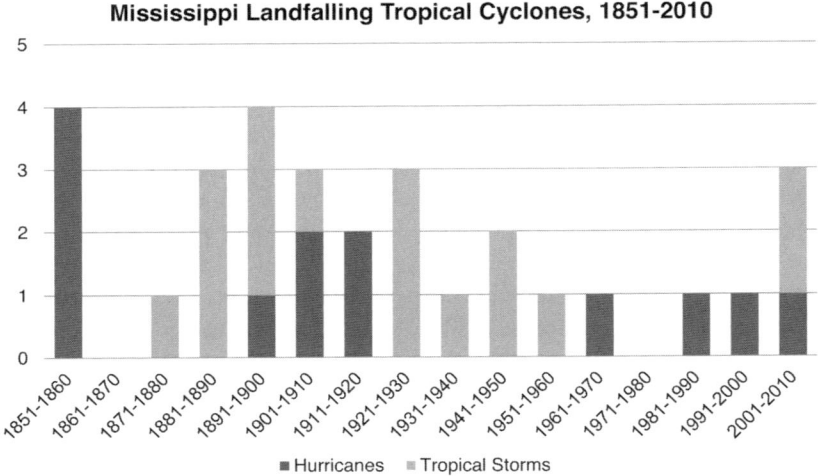

Fig. 6.5. Decadal variation in tropical cyclone frequency, 1851–2010. Based on National Oceanic and Atmospheric Administration best-track data set

north and south of the equator. (Mississippi's coastal counties are located north of 30°N.)

More importantly, hurricanes also need some trigger to begin their development. Tropical cyclones can be initiated by favorable conditions that occur in easterly waves, in the convergence of winds around an area called the Intertropical Convergence Zone, in storms moving from West Africa into the Atlantic, and along old frontal boundaries. Most hurricanes that make landfall in Mississippi do not originate in the Gulf of Mexico, but instead enter the Gulf after forming elsewhere. Those that form in the northern Gulf often do not have enough time to develop into a strong hurricane before making landfall.

Many hurricanes begin as easterly waves (also known as tropical waves), which are troughs of low pressure moving across the Atlantic. Because they are moving from east to west, they look like small ridges on the tropical surface weather map. But these ridges are actually inverted troughs. Hurricanes Camille, Dennis, Gustav, and Katrina all formed from tropical waves, some assisted by other weather features, such as areas of enhanced convection or storminess. Many easterly waves begin off the western coast of Africa and travel all the way across the tropical Atlantic. Hurricane Ivan, which made landfall near Gulf Shores, Alabama, formed from a wave that initiated off of Africa. It became a hurricane before entering the Gulf. Hurricanes Camille and Gustav formed from tropical waves in the Caribbean Sea.

Tropical Storm Allison, which crossed through Mississippi, formed when a cluster of thunderstorms moved offshore of the eastern coast of Mexico.

Hurricane Alicia did not make landfall in Mississippi, but it did form in the northern Gulf. Alicia formed on an old frontal boundary before making landfall as a hurricane in the Galveston and Houston areas. Hurricane Katrina's development was aided by the remnants of a tropical depression and a trough in the upper level of the atmosphere.

Areas of disturbed weather, such as clusters of thunderstorms that move offshore, can help to initiate the development of a tropical system at any time, but there is some seasonality to the other methods. Hurricanes and tropical storms that develop on old frontal boundaries typically form during the beginning or end of the hurricane season. These storms have a better chance of impacting Mississippi at the beginning of the season. Tropical systems that form near old frontal boundaries at the end of the season tend to form off the East Coast. In the height of hurricane season in August and September, fronts do not make it far enough south. During August and September, storms that begin as tropical waves off the African continent are more common.

HISTORICAL HURRICANES

Between 1851 and 2010, there were 13 hurricanes that made landfall in Mississippi. A landfall is marked by the hurricane or tropical storm moving from the water to land in Mississippi. This excludes storms that make landfall in Alabama or Louisiana and then cross into one of the coastal counties, even though some of these storms have caused damage in Mississippi. Figure 6.5 shows the frequency of hurricanes and tropical storms that made landfall in Mississippi from 1851 to 2010.

Although we have records of hurricanes making landfall in the northern Gulf since the time the Spanish first explored the Mississippi River in the early 16th century, the best hurricane records exist from 1950 to the present. The first intentional flight into a hurricane was made in 1943, and after World War II the government regularly sent military planes into hurricanes. The National Hurricane Research Program, in which missions were flown to collect data about the hurricanes, began in 1955, and weather satellites contributed to hurricane research starting in the 1960s (NOAA 2007). The most significant technological advance for hurricane tracking was the use of satellites (Elsner and Kara 1999). Before the use of satellites and reconnaissance aircraft, people had to rely on ship reports to know whether there was a hurricane in the open ocean. When a hurricane made landfall or when a ship crossed a hurricane's path, evidence of the hurricane became part of the historical record. Many early records of hurricanes also come from damage to ships along the coast.

During the 18th century, several storms impacted the Mississippi coast. On September 22–24, 1722, a hurricane made landfall west of the mouth of the

Fig. 6.6. Satellite images of Hurricanes Camille (1969) and Katrina (2005) making landfall in nearly the same location. Image credit: National Climatic Data Center, National Oceanic and Atmospheric Administration, Department of Commerce

Mississippi River. This places the landfall in Louisiana, but damage was report-ed in Biloxi. A hurricane that made landfall at the mouth of the Mississippi River was said to have affected the Gulf Coast from Louisiana to Pensacola. A hurricane of unknown intensity hit the Louisiana, Mississippi, and Alabama area in 1746. Several other storms struck the New Orleans area from the 1770s to the 1790s. It is not clear whether they had any impact on Mississippi, though it is likely at least some of them would have.

The 1772 hurricane was documented by Bernard Romans, a Dutch surveyor, cartographer, and naturalist in his 1775 book *A Concise Natural History of East and West Florida*. Romans stated that the most intense part of the hurricane was felt by people on the Pasca Oocolo River, where the plantation of Mr. Krebs was almost completely destroyed. Krebs lost several buildings and many nearby trees were blown down or broke, according to Romans. The La Pointe–Krebs house, also called Old Spanish Fort, is located in what is now Pascagoula.

A greater number of tropical cyclones were documented in Louisiana, Mississippi, and Alabama in the 19th century. The record begins with a "very destructive" hurricane hitting the Gulf Coast on August 19, 1813, according to meteorologists Gordon Dunn and Banner Miller (1960). Not much is known about this hurricane, including where exactly it made landfall or how much damage was done. A storm on July 27–28, 1819, was also very destructive to the Gulf Coast of Mississippi. The hurricane made landfall in Bay St. Louis and af-fected the whole length of Mississippi's coast. Several ships were wrecked and their crews lost during the hurricane. The storm also caused destruction in Bi-loxi. Reports recounted by Ray Bellande in the Ocean Springs Archives sug-gested that a schooner may have sailed right over the Biloxi Peninsula into the Back Bay due to the storm surge. A similar storm also passed over Bay St. Louis in mid-September 1821. According to the Hancock County Historical Society,

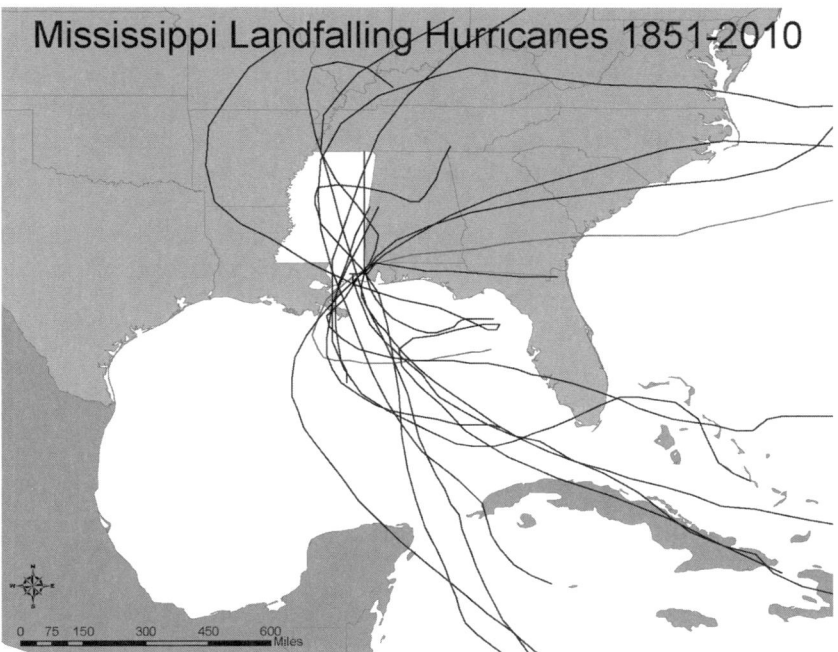

Fig. 6.7. Paths of all hurricanes that made landfall in Mississippi, 1851–2010. Based on National Oceanic and Atmospheric Administration best-track data set

the packet *Washington* and all its passengers were taken by the storm, much the same as had happened in the 1819 hurricane. This storm was described by Dunn and Miller (1960), as minor. The August 1831 hurricane was felt by Mississippi after having made landfall near Baton Rouge. This storm was destructive to New Orleans, but was not as strong in Mississippi. Another hurricane to make landfall in Louisiana, as well as Texas, before moving into Mississippi was the Racer's Storm of October 1837. According to *Louisiana Hurricane History* (Roth 2010), this storm washed away all the wharves along the Mississippi coast. It also claimed many boats, including four steamboats, and much of the sugar cane and cotton crops for that year.

Better hurricane records exist from 1851 to the present. Figure 6.7 shows all the hurricanes to have made landfall in Mississippi from 1851 to 2010. Researchers have gone back through the records and plotted tracks, strengths, and dates associated with these storms. According to this hurricane reanalysis project, four hurricanes made landfall in Mississippi from 1851 to 1860. The August 1852 hurricane made landfall somewhere between Pascagoula and Mobile and was considered minor. The hurricane of September 15–16, 1855, which made landfall near Gulfport as a category 3, was much worse than the 1852 storm. It was

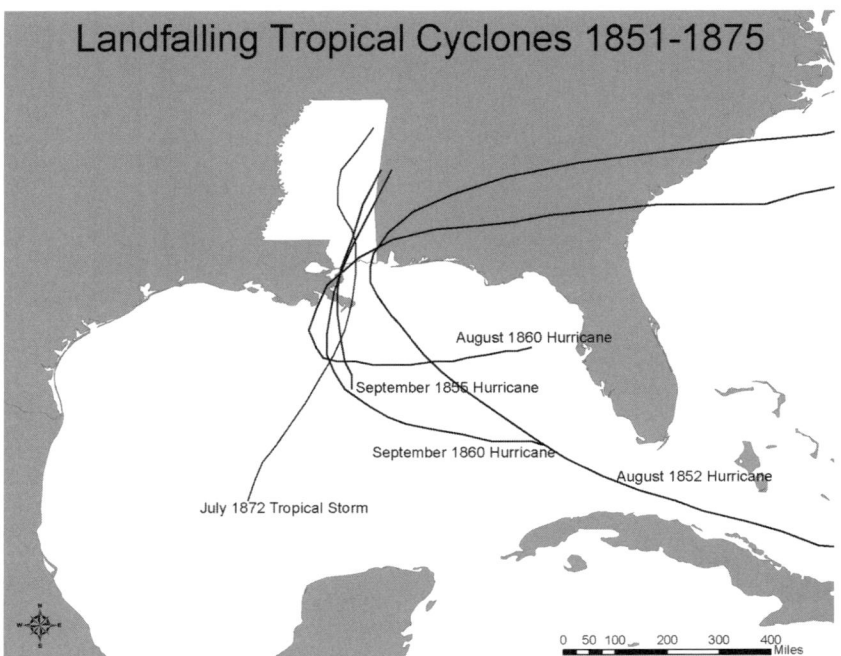

Fig. 6.8. Paths of all hurricanes and tropical storms that made landfall in Mississippi, 1851–1875. Based on National Oceanic and Atmospheric Administration best-track data set

compared with the 1819 hurricane, which was very destructive in Mississippi. Roth (2010) reported it to be a "compact" storm, but most structures on the Mississippi coast were swept away by it. This hurricane destroyed a wharf and pier in Ocean Springs. In fact, there were reportedly only two wharves that survived between Bay St. Louis and Pascagoula (Hancock County Historical Society). Three storms affected Mississippi in 1860, two making landfall here. The first hurricane made landfall on August 11 between Biloxi and Pascagoula, and the second made landfall on September 14–15 over Bay St. Louis. Both storms were considered minor by Dunn and Miller (1960), but even weaker hurricanes can do damage. The August 11 hurricane was said to have drowned 300 cattle on Cat Island and destroyed the tower there (Roth 2010). The September 14–15 hurricane destroyed the lighthouse in Bay St. Louis as well as one of its hotels. The third hurricane in 1860 did not make landfall in Mississippi, but high winds were felt in Natchez.

After the 1860 season, Mississippi experienced an extended calm period. No hurricanes made landfall until 1893, although there were several tropical storms that made landfall and a few hurricanes affected the state. The September 18, 1875, hurricane crossed through central Mississippi, and the August 1888

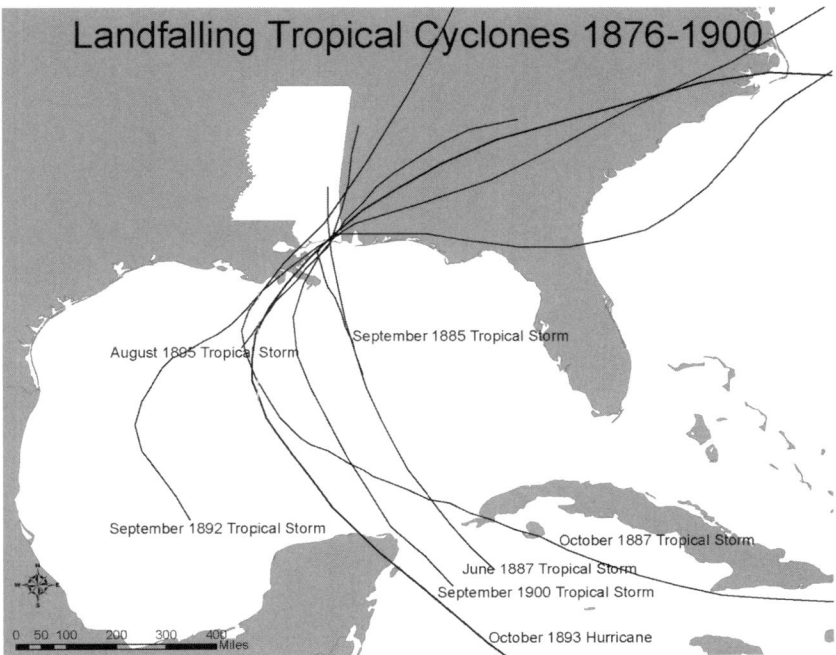

Fig. 6.9. Paths of all hurricanes and tropical storms that made landfall in Mississippi, 1876–1900. Based on National Oceanic and Atmospheric Administration best-track data set

hurricane skimmed the northwest corner of the state. Hurricanes in September 1889 and September 1893 tracked through or near Mississippi, but made landfall east and west of the state, respectively.

The October 2, 1893, hurricane brought far more damage. Called a storm of "great violence" by Roth (2010) and of "extreme" intensity by Dunn and Miller (1960), the October 1893 storm caused a great deal of death and destruction along the Gulf Coast. This storm is sometime referred to as the Chenier Caminanda (or Caminada) Hurricane or the Great October Storm. This hurricane made its first U.S. landfall on October 2 south of New Orleans and west of Grand Isle. After the eye moved back over the Gulf for several hours, it made its second U.S. landfall on the Mississippi coast just west of the Alabama border. As far away as Bay St. Louis, streets were impassible right after the storm. The railroad bridge between Bay St. Louis and Henderson Point was swept away, as was the bridge between Biloxi and Ocean Springs. A factory wharf, the Gulf Coast Market building, the U.S. Marine Hospital, and many smaller buildings succumbed to the storm (information from the *Daily Picayune* on the Hancock County Historical Society website). This category 4 hurricane ranks as the fourth deadliest in U.S. history, with many of the deaths occurring in Louisiana.

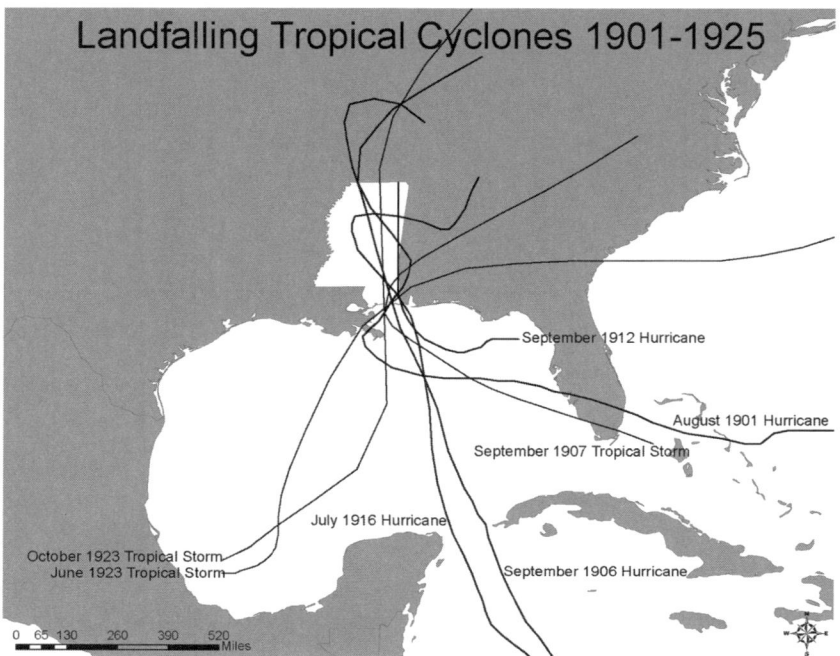

Fig. 6.10. Paths of all hurricanes and tropical storms that made landfall in Mississippi, 1901–1925. Based on National Oceanic and Atmospheric Administration best-track data set

Several tropical storms affected Mississippi after the 1893 hurricane. The next hurricane, a category 1, made landfall on August 15, 1901, southeast of Ocean Springs. According to the Ocean Springs archive, portions of the shoreline were badly damaged and the New Beach Road was swept into the bay. This hurricane also caused 10 deaths. A category 2 hurricane made landfall near the mouth of the Pascagoula River on September 27, 1906. This hurricane brought a lot of damage to agriculture in Ocean Springs; destroyed the 1878 Baptist Church, the Knights of Pythias Hall, and the Horn Island Lighthouse; and damaged many residences. The 1906 storm also caused high storm tides. Even locations far inland, such as Brookhaven and Waynesboro, saw damage. About 10% of that area's virgin timber was destroyed. Although 1909 had an active hurricane season, no hurricane made landfall in Mississippi. Lives were lost in both Mississippi and Louisiana by a strong storm that made landfall about 50 miles west of New Orleans. A weak hurricane skirted the eastern edge of Jackson County in 1912, causing minor damage.

The hurricane of July 5–7, 1916, called the Middle Gulf Coast hurricane, made landfall on the Mississippi-Alabama border as a category 3 storm. The storm then tracked around the state in an interesting fashion, first moving inland near

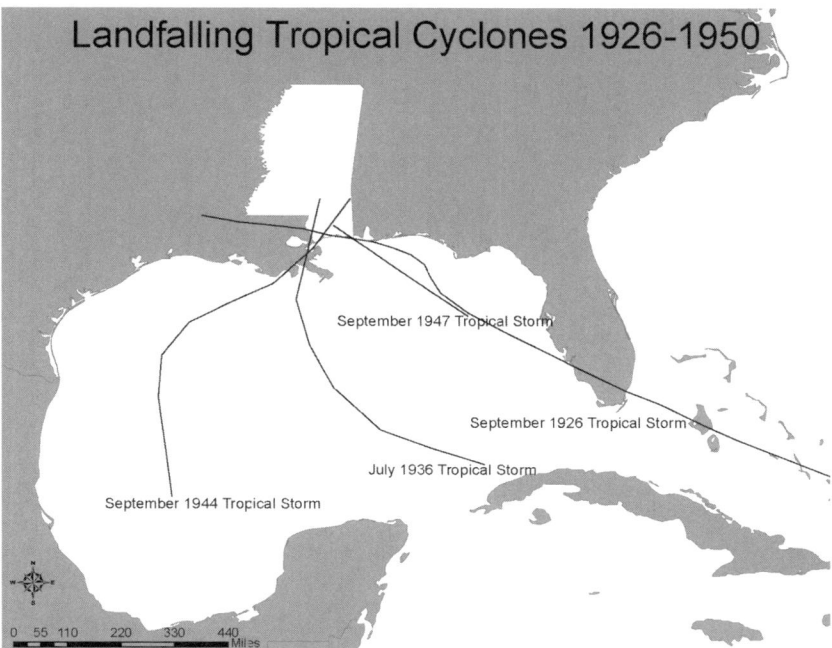

Fig. 6.11. Paths of all hurricanes and tropical storms that made landfall in Mississippi, 1926–1950. Based on National Oceanic and Atmospheric Administration best-track data set

Pascagoula late on July 5 on a northwest track. It then traveled through Jackson to Cleveland, where it turned east during the night, moving over Macon to near Selma, Alabama, where a turn west then carried it over Birmingham and Huntsville on the July 7 and 8. Another turn to the west took it past Nashville and into the Ohio Valley on July 10. The 3 days of heavy rains caused large losses of staple crops and resulted in great floods on the rivers of eastern Mississippi, Alabama, and Georgia. This storm also caused a storm surge of 11.6 feet in Mobile, a record that remains unsurpassed.

Several hurricanes and tropical storms affected Mississippi from the 1920s to the 1960s. A hurricane came very close to making landfall in Mississippi in 1932, moving into Jackson County after making landfall in Bayou la Batre, Alabama. The hurricane of 1947 brought tides of 12 feet to Biloxi, Bay St. Louis, and Gulfport before making landfall in Louisiana. These tides remain the third-highest tide recorded in Pascagoula, Biloxi, Gulfport, Long Beach, Bay St. Louis, and Pass Christian. Although the hurricane did not make landfall in Mississippi, 22 Mississippians lost their lives during this hurricane (National Oceanic and Atmospheric Administration, National Weather Service, and National Hurricane Center, 1993).

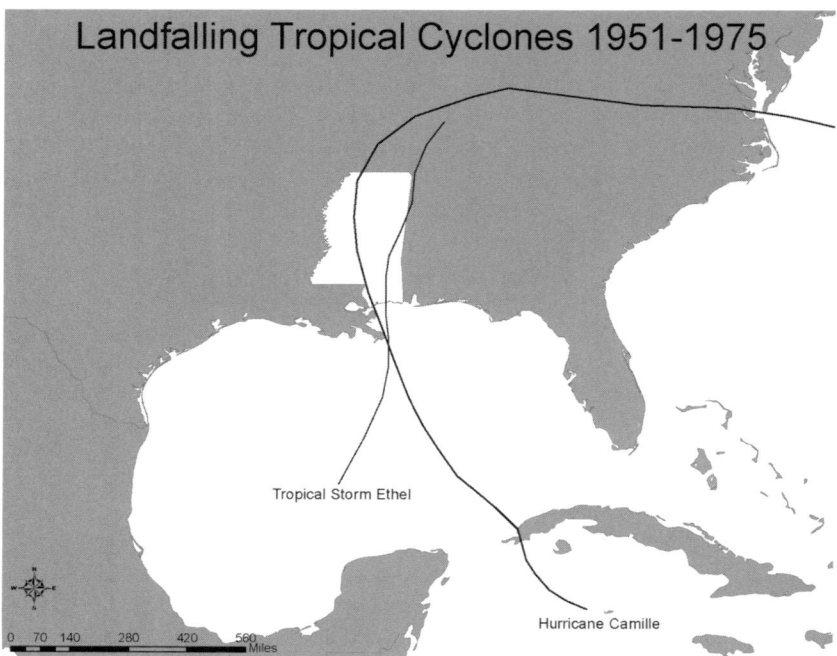

Fig. 6.12. Paths of all hurricanes and tropical storms that made landfall in Mississippi, 1951–1975. Based on National Oceanic and Atmospheric Administration best-track data set

Nonetheless, after the 1916 hurricane, no hurricane made landfall in Mississippi until 1969. By that point, hurricanes were given women's names, and the lady that visited Mississippi in August 1969 was Hurricane Camille. One of only three category 5 hurricanes to make landfall in the United States (the other two are the Labor Day Hurricane of 1935 and Hurricane Andrew), Camille set the standard by which all other hurricanes would be measured along the Mississippi coast for decades to come. Camille originated on August 5 from a tropical wave that moved off Africa, and it became a hurricane on August 15. The hurricane intensified to a category 5 soon after entering the Gulf of Mexico. At the time, Camille's central barometric pressure (26.61 inches) was lower (indicating a very intense storm) than any other U.S. hurricane except the 1935 Labor Day Hurricane in Florida. Hurricane Camille made landfall near Bay St. Louis and Waveland at 11:30 p.m. on August 17, 1969, with both ferocious wind speeds and a high storm surge. Biloxi reported a gust of 229 mph during Camille. The night before the storm hit, one local radio station cautioned listeners that the winds would be "of nuclear force" (as remembered by Irma Reinike in the book *Miss Camille 1969: The Devastating Female*). Because Camille crossed the coastline directly northward with the eye on the western side of the coast, the impact of

Fig. 6.13. Damage from Hurricane Camille in Biloxi, Mississippi. Photo credit: National Oceanic and Atmospheric Administration, Department of Commerce

its storm surge was amplified. (The northeast quadrant of a tropical cyclone has the strongest winds and these are onshore winds, so the storm surge is greatest in that vicinity.) Camille's storm surge was listed as 24.6 feet at Pass Christian in the SUREGDAT database, based on reports by the National Weather Service (NWS) and the Atlantic Oceanographic and Meteorological Laboratory, although other sources state a maximum surge of 22.6 feet (http://surge.srcc. lsu.edu/). The high surge was not confined to Pass Christian. The storm surge was in excess of 20 feet in Bay St. Louis, Long Beach, and Gulfport, and it was nearly 20 feet in Biloxi. Hurricane Camille's storm surge, coupled with the wind, caused massive devastation along the coast. The degree of damage and destruction close to the center at landfall was described by a NOAA post-storm report as "virtually complete" and "resembling more the effect of a tornado than a hurricane." Because Camille retained its hurricane force winds north of Columbia and Hattiesburg, 26 counties in southern Mississippi were declared a disaster area. Hurricane-force wind gusts were felt in Jackson.

Along the coast, boats were washed into yards where homes no longer stood. Many homes and buildings were severely damaged in Ocean Springs and Biloxi. A barge was pushed onto Highway 90 in Gulfport, and many buildings along Highway 90 were completely destroyed. Gulfport Little Theater was reduced to its frame. The Long Beach Shopping Center was severely damaged. Cars and car parts, pieces of houses, trees, and people's belongings littered the streets in Long Beach. Houses of worship such as St. Thomas Catholic Church

Fig. 6.14. Trinity Episcopal Church in Pass Christian (a) before and (b) after Hurricane Camille. Photo credit: National Oceanic and Atmospheric Administration, Department of Commerce

in Long Beach and Trinity Episcopal Church in Pass Christian were destroyed. The Dixie White House in Pass Christian was too badly damaged to save and other parts of Scenic Drive were washed away. All that remained of the Riche- lieu Apartments, where 32 people had remained during the hurricane, was the foundation and debris (Figure 6.15); all but two died. The destruction was

Fig. 6.15. The Richelieu Apartments in Pass Christian (a) before and (b) after Hurricane Camille. Photo credits: National Oceanic and Atmospheric Administration, Department of Commerce

severe in downtown Waveland as well. Camille killed 135 people in Mississippi and caused $1.42 billion in damage (not adjusted for inflation). After leaving Mississippi, Camille killed an additional 113 people from flooding in Virginia. The total death toll from Camille is 256 lives.

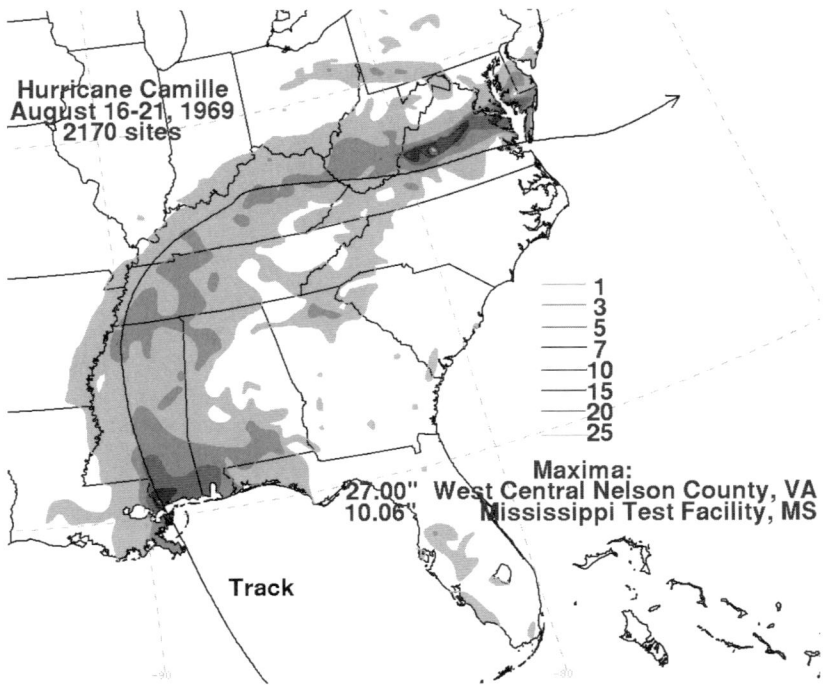

Fig. 6.16. Rainfall during Hurricane Camille. Parts of central Virginia saw 25 inches. Image credit: Hydrometeorological Prediction Center, National Oceanic and Atmospheric Administration, Department of Commerce

After Hurricane Camille, Mississippi experienced another long period with few hurricane or tropical storm landfalls. In July 1979, Hurricane Bob made landfall in Louisiana and then crossed almost the whole length of Mississippi, bringing rainfall, a few tornadoes, and some damage to the coast. Also in 1979, category 3 Hurricane Frederic made landfall just east of Mississippi on Dauphin Island, Alabama. Pascagoula was under the hurricane warning area for Frederic and the eastern Mississippi Gulf Coast experienced high tides, although not as high as in Camille or the hurricane of 1947. Hurricane Elena made landfall in Biloxi on September 2, 1985. Like Camille, Elena also originated as a wave moving off the African continent. It became a hurricane once it entered the Gulf. Elena moved around the Gulf, at one time threatening the Florida panhandle, before changing direction and hitting Mississippi. According to NOAA's "Memorable Gulf Coast Hurricanes of the 20th Century," almost 1 million people from Louisiana to western Florida were asked to evacuate before Elena.

Once again, over a decade passed with no hurricanes. The Gulf Coast of Mississippi had completely rebuilt from the damage of past hurricanes, notably Camille, and in 1992 the state legislature gave permission for casinos to locate

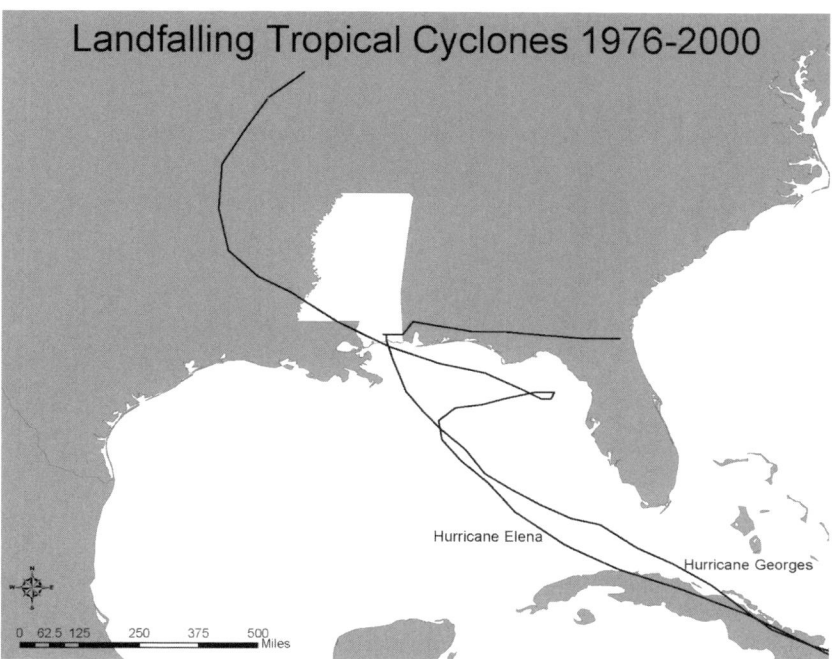

Fig. 6.17. Paths of all hurricanes and tropical storms that made landfall in Mississippi, 1976–2000. Based on National Oceanic and Atmospheric Administration best-track data set

there. The condition was that the actual portion of the casino where the gaming took place had to be located on the water. Much new development took place along with the new casino industry, but the barges they were located on were not tested until 1998, when Hurricane Georges made landfall. Georges was another hurricane that traveled across the Atlantic and moved into the Gulf. After looking like it might make landfall in New Orleans, Georges made landfall near Biloxi as a category 2 hurricane. The casinos sustained damage, especially the Treasure Bay Casino, which suffered major damage to its mooring system. But for the most part, the hotels and other beachfront buildings saw only minor damage. Treasure Bay was able to reopen after 12 days.

Some hurricane researchers have said that the United States entered a more active period of hurricane hits in 1995. In the early 21st century, Mississippi experienced three tropical storms (including one that was actually classified a subtropical storm), including Allison (2001), Hanna (2002), and Cindy (2005). Making landfall in Louisiana and crossing into Mississippi, Allison's damage was caused primarily by heavy rain, although one tornado was reported in George County, Mississippi. The storm caused damage to several manufactured

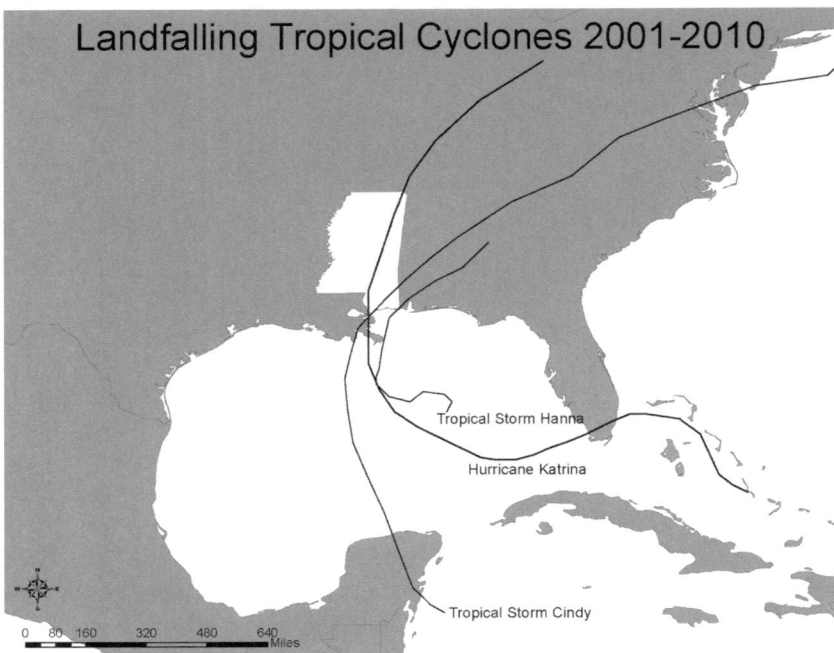

Fig. 6.18. Paths of all hurricanes and tropical storms that made landfall in Mississippi, 2001–2010. Based on National Oceanic and Atmospheric Administration best-track data set

homes, injured the occupant of one mobile home, and killed one person in Mississippi. None of the other tropical storms caused a great deal of damage in Mississippi. The 2005 season experienced the most hurricanes on record, with 27 named storms. Forecasters went through the whole alphabet of names and began to use Greek letters. The 2005 season experienced 15 hurricanes, and 12 tropical storms before ending in January 2006 with the dissipation of Tropical Storm Zeta, which formed on December 30—a full month after the end of hurricane season. The 2005 storm of the most importance to Mississippi, however, was Hurricane Katrina.

After 1969, all hurricanes to threaten the Mississippi coast were compared to Camille, a category 5 hurricane. If buildings survived Camille, then they could withstand any weaker hurricane—or so people all along the coast thought before Hurricane Katrina arrived. On August 29, 2005, Hurricane Katrina made landfall in southeastern Louisiana near Buras as a category 3 storm with sustained winds of 127 mph. The hurricane made a second landfall at the mouth of the Pearl River near the Mississippi-Louisiana border. What was worse than the winds, however, were the waves. All previous storm tide records were broken by

Fig. 6.19. Extent of destruction in Biloxi, Mississippi, following Hurricane Katrina. Very little remained standing within one block of the beach. Photo credit: Commander Mark Moran, National Oceanic and Atmospheric Administration Corps, NMAO/AOC, National Oceanic and Atmospheric Administration, Department of Commerce

Hurricane Katrina. Most places along the Mississippi coast experienced tides of more than 20 feet. The highest reported tide was 27.8 feet in Pass Christian.

The storm brought nearly incomprehensible devastation and loss of life. The four-lane bridge connecting Ocean Springs to Biloxi was swept away, not to be reopened until 2008. The damage grew progressively worse along the coast from east to west. West of Biloxi, buildings several blocks inland were wiped away in many locations. Many of the grand historic homes along the beach-front, which had survived Hurricane Camille, were destroyed. Beauvoir, the home of Jefferson Davis, was badly damaged. Even many of the majestic live oaks that had withstood 200 previous hurricane seasons were felled by Katrina. Those remaining were stripped eerily bare. Hurricane Camille had also caused a powerful storm surge, and all along the coast Camille's tides hold the number two spot after Katrina. Despite being a lower category storm, Hurricane Katrina was a much larger hurricane than Camille, which caused its devastation to be so widespread.

Many people did not leave their homes before Katrina made landfall. Many thought the storm would not be as bad as Camille, and others simply did not want to leave their homes. Those who did evacuate—and the majority of those who were asked to evacuate did—could not have imagined what little there would be to come home to. Katrina caused $81 billion in damage. More than 230 Mississippians and nearly 1100 Louisianans lost their lives in the storm. Most of the deaths in Mississippi were caused by the storm surge, and the death toll was the largest in Mississippi history. Governor Haley Barbour promised the coast would be built back better than before. As this book is being written portions of the coast still have not been rebuilt, but over time it will be. Although Hurricane Katrina was the last hurricane to make landfall in Mississippi, history tells us there will be other landfalls in the future.

HURRICANE IMPACTS

The first aspect of a hurricane that normally comes to mind is wind, and wind is the characteristic that we use to rank them with the Saffir-Simpson Hurricane Wind Scale. In a hurricane, the strongest winds are in the eye wall. The eye is the central area of calm in a hurricane, whereas the eye wall is the most intense portion of the hurricane surrounding the eye. The strongest winds in a hurricane are typically at approximately 1600 feet above ground level. Winds are also stronger to the right of the hurricane track, because the air circulates around the hurricane in a counterclockwise motion. The forward speed of the storm system gets added to the wind speed on the right of the track, while the forward speed is subtracted from the wind speed on the left of the track. For instance, if a hurricane is moving forward at 15 mph, and its wind speed is 100 mph, the winds on the right of the track will be 115 mph. Winds to the left of the track will be 85 mph, because they are blowing against the forward direction of the hurricane and will be diminished by the forward speed. Tropical cyclones that make landfall east of Mississippi may affect the state only slightly because of less intense and offshore winds on the western side of the eye.

Wind can do a great deal of damage in strong hurricanes. At the low end, category 1 hurricane-force winds will blow down branches or signs and cause damage from flying debris. Category 5 winds over 155 mph may cause roof failure or the total destruction of site-built houses. Most insurance policies against hurricane damage only cover wind, but not water, damage during a hurricane. The water damage can be more severe as well as more dangerous.

The storm surge is the rise of water above normal tide levels that accompanies a hurricane. The height of the surge is related to both the wind speed and low pressure of a hurricane. It's a common misconception that the bulge of water called a storm surge is the result of the very low pressures at the center of the hurricane. While this is a component of storm surge, the main cause is the constant drag of the winds over the water. Because winds are the driving force, areas to the right of the track at landfall (which is east of the track in

Fig. 6.20. Damage to the Treasure Bay Casino remains 8 months after Hurricane Katrina. Photo credit: K. Sherman-Morris

Mississippi) can experience the highest storm surge, other factors being equal. This is due to two reasons. First, winds to the left of the track are not as intense. Second, winds to the left are not usually able to pile up water because they are blowing from over land. The exception to this is that winds blowing over a back bay can cause water to pile up in the bay and flood the adjacent land between the bay and the ocean. The surge from the bay on the left side of the hurricane is typically not as great as the surge to the right of the track from the ocean.

The storm surge is often the most damaging impact of a hurricane. The power of the water slowly weakens the bases of buildings until they fail and are washed away. Storm surge has historically been the primary cause of death in hurricanes. Before the widespread use of satellites, it was possible for a hurricane to make landfall with little warning. When this occurred, the death toll from storm surge was high. In the Galveston hurricane of 1900, approximately 8000 people died, primarily due to storm surge. It is also a cause of many deaths in modern times in less developed countries, such as Bangladesh. Most hurricanes affecting the United States have sufficient warning for coastal residents to get out of the way of the surge. Storm surge can still be a killer in the United States, though. It was the primary cause of death in Mississippi during Katrina.

Emergency managers and hurricane forecasters use the predicted storm surge as the primary way to decide whether to evacuate a place before a hurricane. There are tools to help forecasters and emergency managers determine which areas will flood during a hurricane. Geographical features such as the shape of the coastline and how quickly the ocean floor drops off after leaving the shore help to determine how great storm surge will be. Storm-specific features such as the hurricane's angle of approach, intensity, and forward speed also tell how high the storm surge will be. All the possible combinations of

Fig. 6.21. Damage from Hurricane Katrina's storm surge along Highway 90 in Gulfport, Mississippi. Photo credit: Greg Nordstrom

these features are put into a computer program that makes a map of the worst possible storm surge height for each hurricane category. Once this information is known, evacuation maps can be drawn based on these zones. Mississippi has three evacuation zones. Zone A will flood in a category 1 or 2 hurricane, zone B will flood in a category 3 hurricane, and zone C will only flood in a category 4 or 5 hurricane.

Tornadoes are a common occurrence in hurricanes, because there is a lot of rising air and instability in a hurricane. When a hurricane comes onshore, the difference between the land and water surface causes the wind to slow down near the ground, while wind speeds higher above the ground do not decrease. This causes winds in the hurricane environment to vary across different heights, a factor that helps tornadoes to form during a hurricane. Because conditions are best for tornadoes near the point where the hurricane comes onshore, most of them do not occur too far inland from landfall. About half the hurricane-induced tornadoes in Mississippi have occurred within 30 miles from the coast and within 24 hours of landfall. The majority have also occurred within 300 miles from the hurricane's center in its outer rain bands. They can occur well inland, however. Compared to other tornadoes, hurricane-induced tornadoes tend to be weaker and shorter-lived. They also have a preferred location with

Table 6.2. Highest hurricane storm surges on record in Mississippi			
Location	Katrina	Camille	Hurricane of 1947
Bay St. Louis	25.0 feet	21.7 feet	15.2 feet
Pass Christian	27.8 feet	24.6 feet	13.4 feet
Long Beach	25.7 feet	21.6 feet	14.0 feet
Gulfport	24.5 feet	21.0 feet	14.0 feet
Biloxi	22.0 feet	19.5 feet	11.1 feet
Pascagoula	18.0 feet	11.8 feet	9.0 feet
Data are from the Weather Underground website (http://www.wunderground.com/hurricane/surge_us_records.asp) except for Hurricane Camille, which are from the SURGEDAT database (http://surge.srcc.lsu.edu)			

respect to the hurricane's movement: to the right of the hurricane's forward motion, which is the same general area with the greatest storm surge threat and most intense wind speeds.

The hurricane that brought the greatest number of tornadoes to Mississippi did not make landfall here. It was Hurricane Rita in 2005, which made landfall near the Louisiana-Texas border. Rita caused 55 tornadoes across the Jackson NWS warning area, including one F3 and seven F2 tornadoes. This tornado outbreak caused one fatality in Humphreys County and 16 injuries across the state. Mississippi also saw a large number of tornadoes (26) associated with Hurricane Andrew, which made landfall in Florida and Louisiana in 1992.

The last of the four major impacts from tropical systems is inland flooding. Tropical cyclones can bring a great deal of rainfall with them. This is true whether they make landfall as a hurricane or tropical depression or even if they stay offshore while hugging the coastline. Storms can produce heavy rain hundreds of miles inland as well as along the coast. Flooding is made worse when an area has certain geographic characteristics, such as mountains or river valleys. When an air mass is forced up a mountain, more precipitation typically occurs than if the ground were flat. This is called an orographic effect. Mississippi does not have to worry too much about this, as the area near the coast is fairly flat. The orographic effect is a problem for the Mid-Atlantic states, which change from coastal plain to the Appalachian Mountains in only a few hundred miles. It also occurs in southern Texas, where flooding is a problem when hurricane moisture meets the Balcones Escarpment. Flooding can also occur along river valleys. The threat of flooding is increased when a hurricane travels upstream in a single river valley or over one river's watershed rather than crossing several different river valleys. The abundance of rainfall, sometimes in excess of 10 inches in a single day, can also be too much for city streets and drainage systems to handle, causing localized flooding.

SEASONAL HURRICANE PREDICTION

Each year different teams of forecasters issue a hurricane forecast for the upcoming season. The most popular one among the mass media and therefore the public is the one issued by Colorado State University forecasters Philip Klotzbach and William Gray. Klotzbach and Gray use various climatological features to predict whether the upcoming season is going to be above or below normal and by how much. Some of the predictors they have used include sea surface temperature at different times of the year, heights of specific levels of the atmosphere in the North Atlantic, and sea level pressure over the tropics. They take into account the weather cycles such as El Niño/La Niña that are known to have an impact on tropical cyclone frequency. They also look at the amount of hurricane activity in previous seasons when the climatic conditions were similar. Unlike Klotzbach and Gray, who issue a prediction of a specific number of hurricanes, National Oceanic and Atmospheric Administration issues an outlook with information about how likely the upcoming season is to be above or below normal.

For the 2010 hurricane season, Klotzbach and Gray predicted a season with a higher than average number of hurricanes and named storms. (The average is about 6 hurricanes and 10 named storms.) They predicted this level of activity because of the development of La Niña conditions, which usually lead to more tropical cyclones, warmer than normal sea surface temperatures in the Tropical and North Atlantic, and lower than usual sea level pressure in the Tropical Atlantic. The Colorado State University forecast team usually issues its forecasts in December, April, and several times during the hurricane season. In the June 2010 forecast, Klotzbach and Gray predicted 18 named storms and 10 hurricanes. There were actually 19 named storms and 12 hurricanes. True to the prediction, 2010 was an active season; however, only one tropical storm and no hurricanes made landfall in the United States.

Both storm surge and inland flooding consist of an excess of water over a land surface. The distinction between the two is that the former is a rise in sea water and the latter is caused by fresh water, either from the buildup of rain or from rivers overtopping their banks. The difference may not seem that important if you are living in an area regularly threatened by hurricanes. To researchers who study impacts of hurricanes on people, however, it could make the difference in how to protect people from harm during a hurricane. For most of history, the storm surge was the top killer during a hurricane, and inland flooding was not given as much importance in forecasting; but many more people have died from inland flooding in modern times. A study by Edward Rappaport (2000) showed that from 1970 to 1999 over 80% of all deaths due to hurricanes were caused by water. The biggest killer (accounting for 59% of deaths) was inland flooding. Although this study did not include Hurricanes Camille or Katrina, it did include some storms of high magnitude, such as category 4 Hurricane Hugo in 1989 and category 5 Hurricane Andrew in 1992. Also, about half the deaths from Hurricane Camille were from inland flooding in Virginia.

As in Hurricane Camille, deaths attributed to inland flooding may not occur in coastal counties. Some people who do not live along the coast may underestimate the danger associated with inland flooding. Most of the deaths caused by non-storm surge flooding could have been prevented by individuals making safer choices. For example, of all the deaths caused by flooding with known locations, 63% were individuals in vehicles (Ashley and Ashley 2008). While some deaths will be unavoidable, in the last decade forecasters have given more emphasis to the risk associated with inland flooding in hurricane warnings.

HURRICANE PREPAREDNESS

It's important to make sure you would know what to do during a hurricane. There are some things you can get ready if you live in an area threatened by the occasional storm.

If you are under an evacuation order, you should leave the area. If you stay, you should plan to be on your own for at least 72 hours. It may be this long until either basic services are restored (if it is not a particularly bad storm) or until relief workers can get into the area. You should have enough food and water to last this long. A gallon of water per person per day is recommended. Batteries will also be important for flashlights and battery-operated radios because power will most likely be out for at least some of the time you are waiting out the storm. Because of the possibility of power outages, the Mississippi Emergency Management Agency (MEMA) recommends turning your refrigerator to maximum cold and opening the door as little as possible. They also recommend picking up any loose items from outside your home and filling the bathtub with water for sanitary purposes. (This is in addition to the amount of drinking water you should keep.)

Whether you stay to shelter in place or evacuate, MEMA recommends assembling a disaster kit. Items to place in your disaster kit include a NOAA weather radio, copies of your important papers (such as social security card, insurance policies, will, medical and tax records, birth and marriage certificates, and proof of residence), cash (because banks may not be open and ATMs may not work or may be out of money), the usual sanitary supplies along with bleach, any extra medications that you may need in the next several days, sunscreen, mosquito repellent, and items to keep you comfortable such as blankets, bedding, and clothing. If you plan to stay, a generator can be helpful during power outages. Gas-powered generators produce carbon monoxide, so they should not be used indoors. Windows and glass doors should be protected with hurricane shutters or plywood. If using plywood, it should be at least 5/8 inch thick, cut to fit inside the window frame, and fastened with lag bolts or barrel bolts. Crossing windows or doors with masking or duct tape does not protect them from winds or flying debris.

Do not forget your pets. While some veterinarians and boarding kennels may accept animals from individuals who are evacuating, remember that they are local residents themselves and may also need to evacuate. Many providers of care for dogs and cats have policies that mandate you must come and pick up your pet once a hurricane warning or evacuation order is issued. Some may impose a hefty fee if you do not. Because many animals died in Hurricane Katrina, more local areas are beginning to provide a shelter for animals during hurricanes as well. During the 2008 Hurricane Gustav, which threatened but did not make landfall in Mississippi, a separate pet facility was opened. Red Cross shelters do not accept pets, and if it is not safe for you in your home during a hurricane, it will not be safe for your dog or cat either. If you decide to evacuate with your pets, many hotels will allow animals to stay with their owners. There are also items that you should have ready for your pet. Your pet emergency kit should have proof of vaccinations and license information, prescriptions or extra refills for any medications your pets take, a leash, and a crate or kennel for your pet to stay in if you need to stay at a hotel. Extra items such as your pet's regular bed can help make the time away from home less traumatic. It is a good idea to get copies of prescriptions for medications because your veterinarian may not be open after landfall if the hurricane is a bad one.

Further Reading

Elsner, J. B., and A. B. Kara. 1999. *Hurricanes of the North Atlantic: Climate and Society.* New York: Oxford University Press.

Keim, B. D., and R. A. Muller. 2009. *Hurricanes of the Gulf of Mexico.* Baton Rouge: Louisiana State University Press.

7. WINTER WEATHER

Fig. 7.1. The January 2000 snow event at Mississippi State University left about 8 inches. Photo credit: L. Lynch

Although summer is the dominant season in Mississippi, winter weather resulting in a significant accumulation of ice and snow is a recurring climatological
trait in the state. Snow is generally minimal and is limited to December, January,
and February, but several large snow and ice events have been recorded in Mississippi. Representative amounts of snow for the east-central part of the state
are 0.1, 0.6, and 0.3 inches for December, January, and February, respectively.
However, there are records of 4, 10, and 11 inches for the same months. These
relatively small amounts of snow result in disproportionately large impacts in
the state, which are exacerbated by a lack of snow and ice removal equipment.
Impacts often include municipal and commercial closures, traffic congestion
and accidents, school closings, shortages of consumables and an increased ner-
vousness about weather among the public because of the disruption of routine.

Legend

Average Annual Snowfall (Inches)

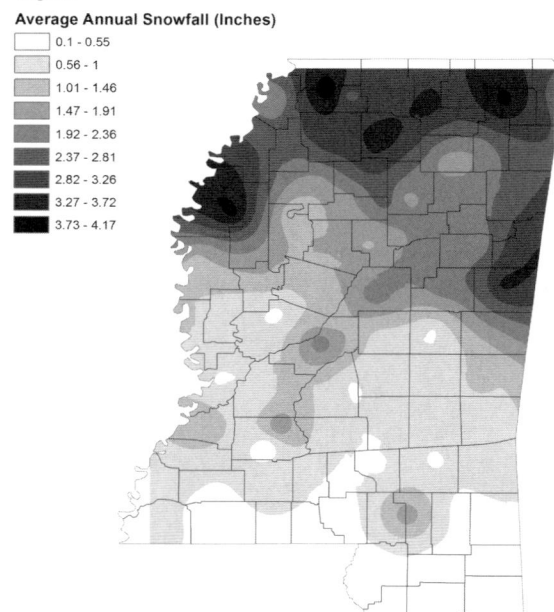

	0.1 - 0.55
	0.56 - 1
	1.01 - 1.46
	1.47 - 1.91
	1.92 - 2.36
	2.37 - 2.81
	2.82 - 3.26
	3.27 - 3.72
	3.73 - 4.17

Fig. 7.2. Average annual snowfall in Mississippi. Data from National Climatic Data Center

Winter storms occur irregularly in Mississippi, but they can be associated with significant costs to citizens and state and local governments. These events are not expected to do widespread structural damage, but they may cause extensive damage to lines of communication and utility services, and debris cleanup costs can be substantial. Schools, government offices, grocery stores, and even critical service providers such as police stations and hospitals may be closed for up to several days due to severe winter weather events across the state.

WINTER WEATHER CLIMATOLOGY

Figure 7.2 shows the average annual amount of snowfall throughout the state. This map may be a bit misleading, however. In a typical winter, some parts of the state may get snow, while other parts may not. Even though the northern tier counties and some of the Delta counties show a yearly average snowfall of 3 inches or more, it does not mean that in every year (or even in most years) these counties will see that much snow. In the South, where snow is not common, the yearly average can be influenced by extremes. For example, when one

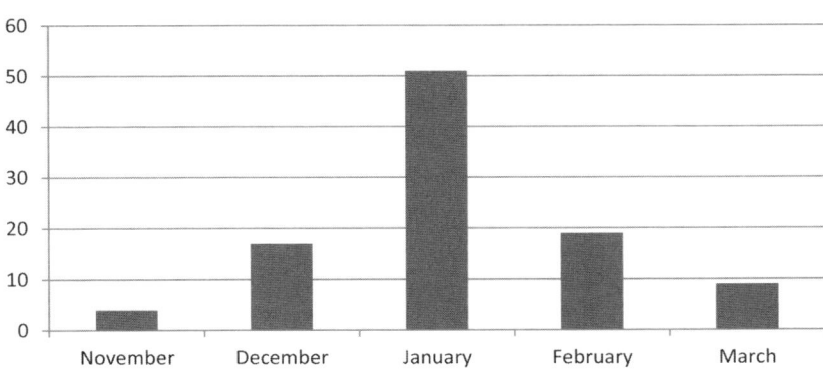

Fig. 7.3. Percentage of snowfall events that occurred in each month in Mississippi, 1961–2001. Data from Duke (2004)

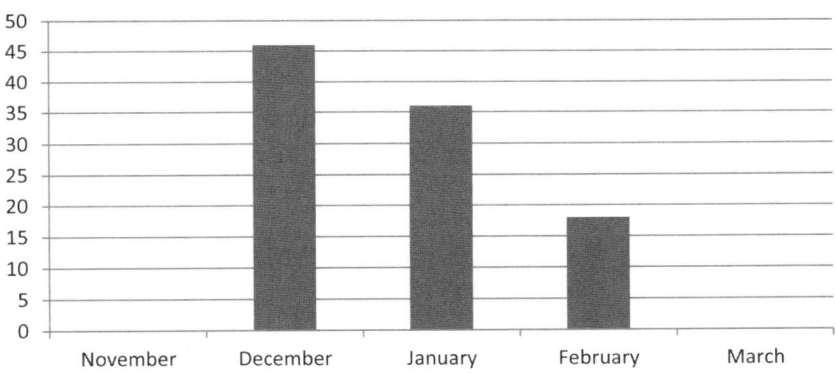

Fig. 7.4. Percentage of ice events that occurred in each month in Mississippi, 1961–2001. Data from Duke (2004)

5-inch snowfall event occurs over a 5-year period, the annual average would be 1 inch. What you can see from the map are the areas in the state where snow is more likely. Generally, snow becomes less common as you move from north to south. But, there does appear to be at least a little more than just north-to-south variation.

When snow does fall, it is most common in January. The data in Figure 7.3 were based on storm data from the National Climatic Data Center and may not reflect every snowfall. Over the 40-year period of 1961–2001, there were 47 snowfall events in Mississippi that were recorded in the storm data (Duke

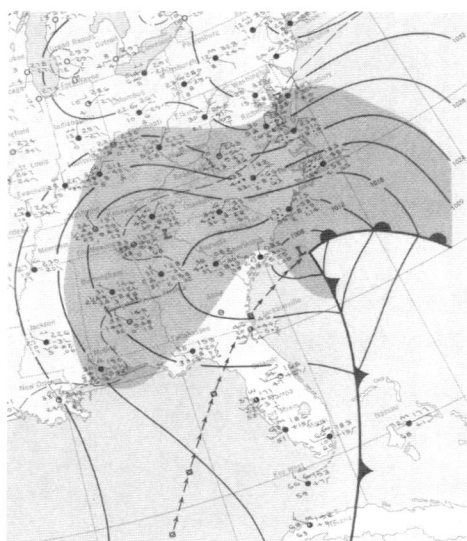

Fig. 7.5. Surface map showing conditions after the New Year's Eve snowstorm of 1963. The surface map is from January 1, 1964, and shows the low pressure system after it had moved to the east of Mississippi. The low pressure system formed in the Gulf of Mexico and moved northeast and along the East Coast. Image credit: National Oceanic and Atmospheric Administration, Department of Commerce

2004). Of these events, 51% occurred in January. The most intense storms also occurred during this month, according to Duke (2004).

In contrast, ice was more common throughout this time period during the month of December. There were only 11 ice events reported in storm data from the National Climatic Data Center (Duke 2004). Throughout the rest of the Southeast, ice events were slightly more common in January, according to this study.

CAUSES

Severe winter weather in Mississippi has some recognizable causes. Upper level lows are generally known to cause snow flurries with little accumulation. When a deep layer of cold air is in place over the state and a surface low develops in the Gulf of Mexico, large accumulations are possible. For example, this scenario caused accumulations of more than 14 inches in some parts of the state in March 1968. Ice storms are caused when precipitation from warmer air aloft falls through a shallow layer of subfreezing air at the surface, freezing immediately upon landing on below-freezing surfaces, such as tree limbs or power lines. Sleet occurs when precipitation from warmer air aloft falls through a deep layer of subfreezing air at the surface, turning raindrops into ice pellets. Whether the layer of air close to the surface is deep or shallow will determine whether the air is cold enough to freeze the water droplets before they reach the ground or when they come in contact with it.

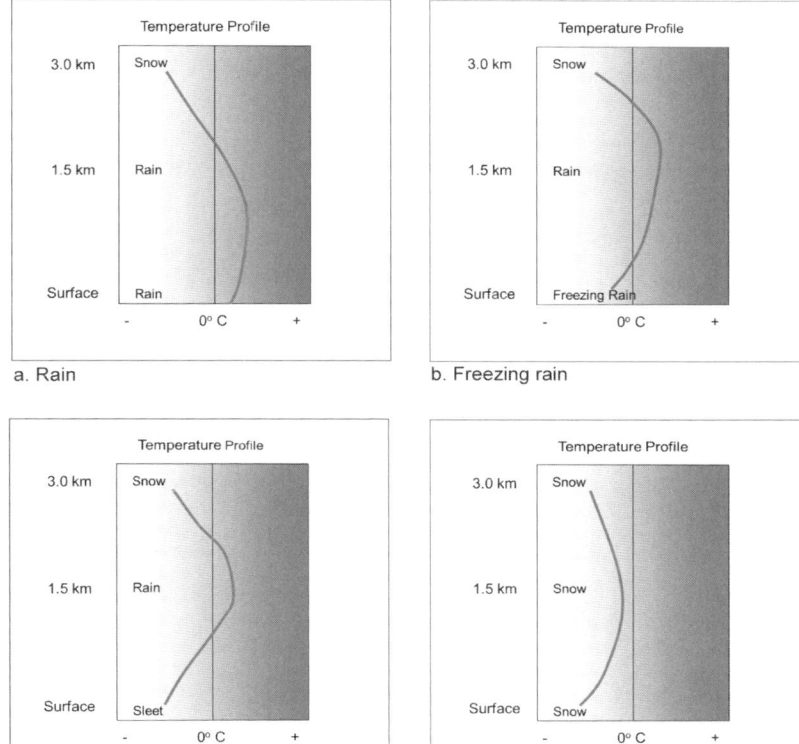

Fig. 7.6. Atmospheric temperature profiles associated with winter precipitation types: (a) rain, (b) freezing rain, (c) sleet, and (d) snow

When determining the type of winter weather a region might experience, one must look at the vertical temperature profile of the atmosphere. Temperatures that remain below freezing for the entire depth of the atmosphere are associated with snow. However, a layer of above-freezing temperatures and its location within the atmosphere will determine if the resultant precipitation will fall in the form of cold rain, sleet, or freezing rain.

Cold fronts moving south through Mississippi tend to stall once reaching the Gulf of Mexico, resulting in the conditions described above. Due to the temperature differences along these fronts, low pressure centers can form in the western Gulf of Mexico. As the low progresses eastward, warmer and moister air is pushed above the colder air at the surface. The moist air aloft provides the precipitation, and the temperature profile of the atmosphere (deep or shallow cold air) determines the precipitation type.

Fig. 7.7. Impacts of the December 1998 ice storm on the Mississippi State campus. Photo credit: University Relations, Mississippi State University

Severe Arctic outbreaks, or cold waves, are sometimes associated with a phenomenon called the Siberian Express, in which bitterly cold and dry air comes into the state from the northern polar region. For example, the cold outbreak around Christmas of 1989 brought temperatures as low as −8°F in the northern part of the state, 4°F in Jackson, and 8°F along the Mississippi coast. High temperatures were in the teens as far south as Jackson and single digits in the northern portion of the state. Many families enjoying Christmas gatherings were without basic plumbing when even drain pipes froze solid. Fortunately, severe Arctic outbreaks are not frequent in the state.

ICE STORMS

During the early morning hours of February 9, 1994, freezing rain began falling over northern Mississippi and continued through midday on February 10. Ice accumulations of 3 to 6 inches were common over the northern and central portions of Mississippi. Due to the weight of the ice, power lines, trees, and tree limbs were downed. Many trees fell on houses and automobiles, and the ice storm caused significant damage to approximately 3.7 million acres of commercial forestland. The damage to urban trees was estimated to be $27 million and that of timber was estimated to be $1.3 billion. About 25% of the state's

Fig. 7.8. Because many trees still had their leaves, there was extensive damage from the ice storm on the Mississippi State campus. Photo credit: University Relations, Mississippi State University

pecan crop was lost for the 5 to 10 years following the event, at an estimated loss of $5.5 million per year. More than 8000 utility poles were pulled down by the weight of the ice, and more than 4700 miles of power lines were downed. Nearly 750,000 customers were without power in the affected area, and some customers were without power for up to a month. In total, 491 water systems were affected, with around 741,000 customers without water. Estimates of damage to utilities run nearly $500 million. This is the worst ice storm to strike Mississippi since a severe ice storm struck the state in January 1951.

Because of its timing close to Christmas, many people also remember the ice storm that began on December 22, 1998 (Figures 7.7 and 7.8). This event was caused by a shallow layer of Arctic air with a layer of warm, moist air above it. Periods of freezing rain and sleet occurred in Arkansas, Mississippi, and Louisiana from December 22 to 25. Many trees had not yet lost all their leaves, and the leaves provided an additional surface for the ice to accumulate on. As much as 2 inches of ice accumulated on trees and power lines. This ice storm caused an estimated $16.6 million in damages.

HEAVY SNOW

A winter storm brought a swath of heavy snow across north-central Mississippi in January 2000. The snow began falling over western portions of the area

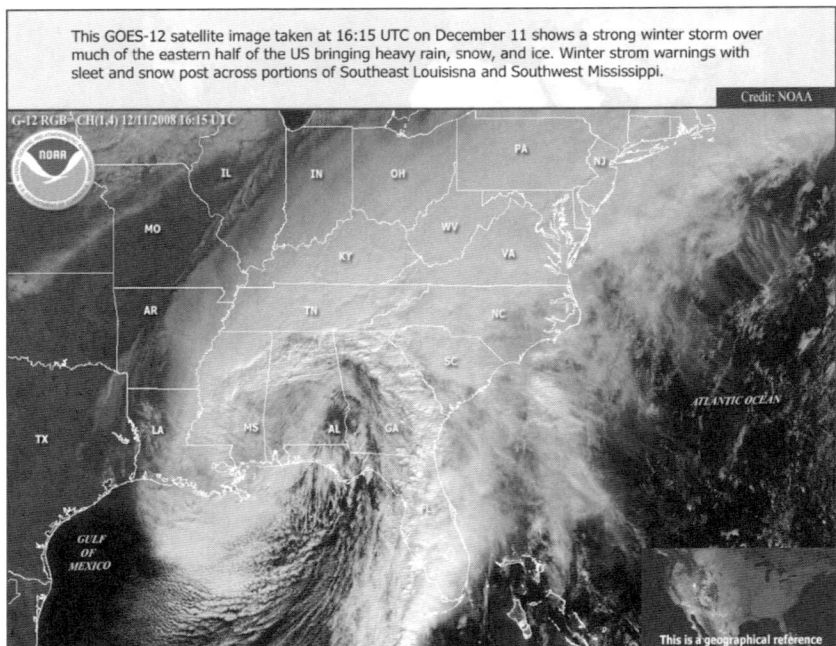

This GOES-12 satellite image taken at 16:15 UTC on December 11 shows a strong winter storm over much of the eastern half of the US bringing heavy rain, snow, and ice. Winter strom warnings with sleet and snow post across portions of Southeast Louisisna and Southwest Mississippi.

Credit: NOAA

Fig. 7.9. A perfectly shaped low pressure system brought the severe weather, rain, and snow that occurred December 9–11, 2008. The satellite image above is from December 11. Image credit: National Oceanic and Atmospheric Administration, Department of Commerce

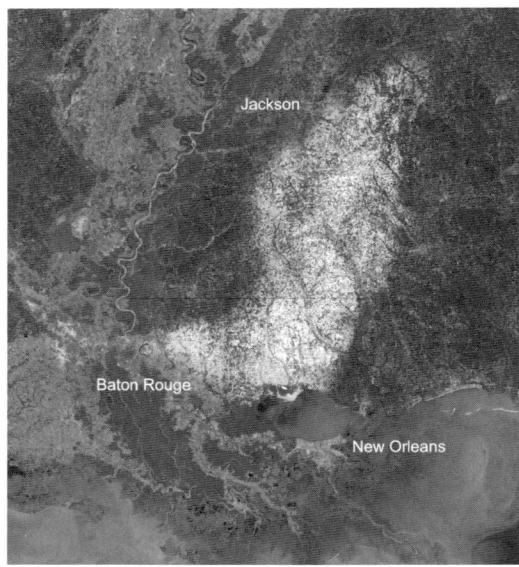

Fig. 7.10. NASA's Terra satellite captured this swath of snow that fell on December 11, 2008. Image credit: National Aeronautics and Space Administration's Earth Observatory

during the early morning of January 27 and spread eastward during the day. The snow was heavy at times and did not end until the morning of January 28. Snowfall amounts generally ranged from 4 to 10 inches. The heaviest amounts fell along the Highway 82 corridor from Greenville to Starkville, where isolated snow depths of 12 inches were reported. Damage from the heavy snow was relatively minimal, with reports limited to a few collapsed roofs and downed trees. Power outages were sporadic, but traveling was more than just an inconvenience, as numerous vehicles ran off the road.

An early and somewhat unusual snow event that took place in December 2008 illustrates the extremes that come with Mississippi winter weather. On December 9, warm air was surging northward ahead of a frontal system. By the late afternoon, the temperature in Jackson had reached 71°F. It was in the upper 60s along the coast and the upper 50s in the northern part of the state. According to the Storm Prediction Center, 146 reports of severe wind, hail, and tornado damage were made in Mississippi that day, including 21 tornado reports.

There was also heavy rain ahead of an approaching cold front. More than 2 inches fell in central Mississippi, with higher amounts in the southern Delta counties. More rain fell on December 10. Cold air entered the state behind the front, and a second low pressure system formed in the Gulf. Many times when this happens the moisture needed to produce snowfall only occurs ahead of the front and then dry air is ushered in with the cold air. This time, however, enough moisture and temperatures just barely cold enough came together to produce a swath of snow, heavy in some locations in southwestern to central Mississippi. Figure 7.10 is a NASA satellite image that shows the distribution of the snowfall. You can tell the swath is snow rather than clouds because you can see the rivers and other surface features.

The snow that fell on December 11 ranged from 9 inches in northwestern Jefferson Davis County to none throughout most of the state. Table 7.1 lists some snowfall amounts from the storm. The amount of snow that accumulated could also have been reduced during this event because of the amount of rain that fell in the days prior. Snow tends to melt quicker when the ground is either wet or not sufficiently cold. One interesting fact about this event is that the temperature was not below freezing in many of the places that saw snow. Many locations' temperatures hovered in the low 30s, from 32 to 35°F. As the heavy snow fell through the atmosphere, some of it melted or evaporated, cooling the air just enough.

WINTER STORMS: A MIXED BAG OF PRECIPITATION

Have you ever heard the weathercaster on television call for a "wintery mix" or a "mixed bag" of winter precipitation? That is because many winter storms do

Table 7.1.	
A sample of snowfall reports from Jackson National Weather Service from December 11, 2008	
Locality	Snowfall
East of New Hebron, Jefferson Davis County	9 inches
Bogue Chito, Lincoln County	8 inches
Martinville, Simpson County	7 inches
Columbia and Kokomo, Marion County	5 inches
Taylorsville and Mize, Smith County	4 inches
Puckett, Rankin County	3 inches
Purvis, Lamar County	2 inches
Sebastapol, Scott County	1 inch
Data from http://www.srh.noaa.gov/jan. Reports were originally from emergency management, law enforcement, post offices, NWS employees, CoCoRaHs and cooperative observers, broadcast media, and the public.	

not come with only one kind of precipitation. Many winter storms in Mississippi are caused by a low pressure system forming near or over the Gulf of Mexico. Moisture is added by the Gulf and surges ahead of a frontal system, sometimes getting wrapped around it. If the air is cold enough and deep enough, the moisture that gets wrapped around a low can become snow. Warmer air can also rise over the cold air, which can set the stage for ice. Figure 7.11 shows a surface map forecast with a common distribution of snow and ice in a winter low pressure system.

The images in Figures 7.12 and 7.13 are from the winter storm that affected northern Mississippi and the Memphis area from January 28 to 30, 2010. During this event, a cold front brought Arctic air south into Mississippi while a low pressure system brought the moisture from the Gulf. Ice accumulations of up to 0.5 inch occurred in northeastern Mississippi. Light snow also swept across the counties on the Tennessee border. As much as 8 inches of snow and 1 inch of ice accumulated between Memphis and Jackson, Tennessee. Corinth, Mississippi, reported 1 inch of snow on top of 0.5 inch of sleet. The National Weather Service also reported that driving conditions on the roadways were treacherous.

THE LATE FREEZE

In addition to the different precipitation-related forms of winter weather, another form of winter weather can have disastrous consequences for Mississippi. This kind of event occurs when winter "lasts too long" or comes back with a vengeance after spring has started—the late freeze. The average dates of last

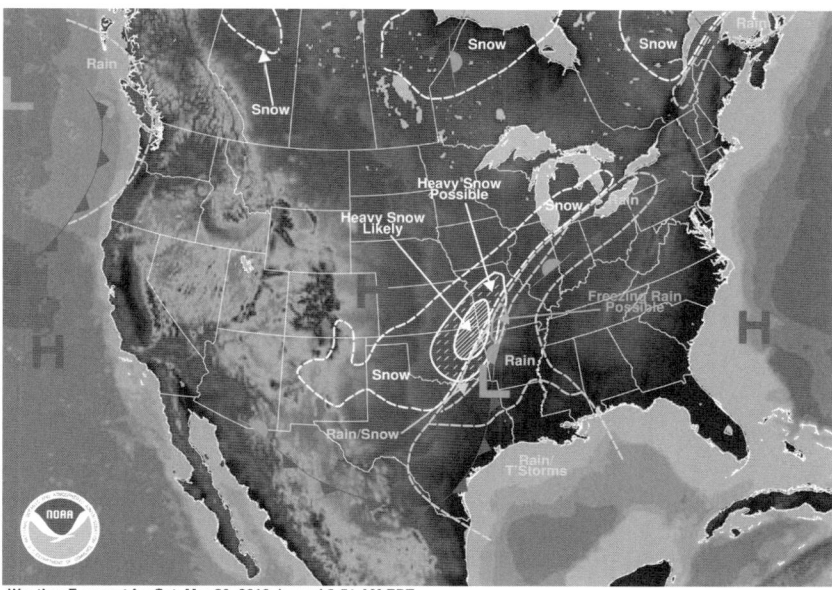

Weather Forecast for Sat, Mar 20, 2010, issued 3:51 AM EDT
DOC/NOAA/NWS/NCEP/Hydrometeorological Prediction Center
Prepared by Hamrick based on HPC, SPC, and TPC forecasts

Fig. 7.11. This weather forecast shows that heavy snow, thunderstorms, rain, and freezing rain were all possible in this late winter storm in 2010. Image credit: National Oceanic and Atmospheric Administration, Department of Commerce

freeze throughout the state show a wide range. For example, the earliest last spring freeze recorded by the weather station at Mississippi State University occurred on February 8, whereas the latest last spring freeze at that location was April 21. This aspect of Mississippi's climate makes planning for agriculture and many outdoor activities very difficult in many years. It can also cause severe economic loss in the state. A notable example was the late freeze of Easter 2007 (Wax 2008).

Impacts of the April 2007 freeze event were both more damaging and, at the same time, less damaging as a result of unusual climate factors preceding the event. The first factor was the abnormally warm March temperatures. Statewide, March temperatures were a little over 5°F above normal. In the northern part of the state, where the freeze event was most pronounced, March temperatures averaged almost 7.5°F above normal. The second factor was that the statewide average rainfall for March was only a little over 1.5 inches, far below normal and the driest March on record in many places in the state. As a result, plant development was somewhat different than would be typical for this time of year, and freeze damage was therefore highly variable. For example, the development of winter wheat was about 2 weeks ahead of schedule, and early heading stages are extremely sensitive to freezing temperatures, so wheat suffered significant

Fig. 7.12. Ice accumulates on some branches in the Memphis area during the storm of late January 2010. Photo credit Mike Goldstein

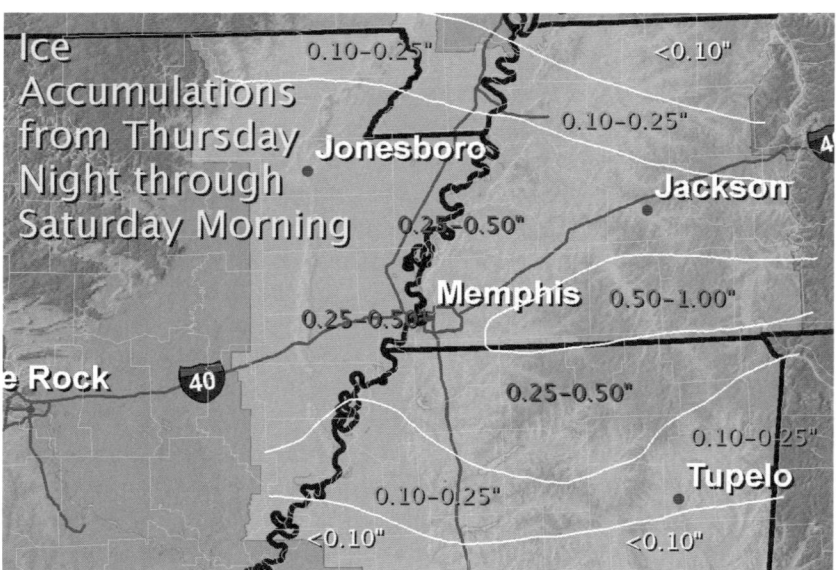

Fig. 7.13. The Memphis National Weather Service made this graphic following the ice storm of January 28–30, 2010. These are ice accumulations, not snow. Only 0.25 inch of ice can damage power lines. Image credit: National Oceanic and Atmospheric Administration, Department of Commerce

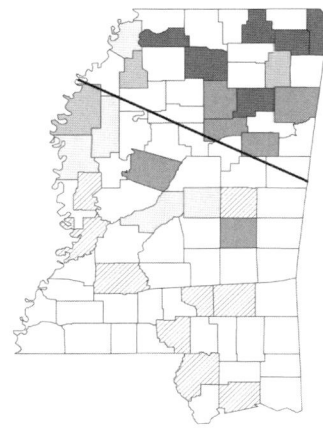

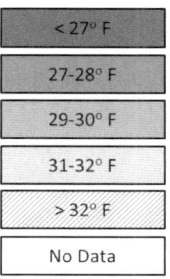

Temperatures
April 7-8, 2007

< 27° F
27-28° F
29-30° F
31-32° F
> 32° F
No Data

Fig. 7.14. Temperatures and damage axis from the late freeze of April 7–8, 2007. Image recreated from data provided by Mississippi Agricultural Statistics Service

damage. However, fruit was also ahead of schedule and was past the growth period during which freezing temperatures are most damaging (March 15–April 1 in Mississippi), so fruit escaped freeze damage.

The coldest period was April 7–8, with temperatures averaging as much as 22.0°F below normal. Freezing temperatures were reported during this 2-day period in the northern third of the state. The lowest temperature was 22°F at Iuka in the northeastern corner of the state and Sardis Dam in the north-central part of the state. Hourly temperature records indicate that the temperature remained at or below 32°F between April 7 and 8 for 12 hours at Tupelo, 6 hours at Memphis (representative of the northwestern corner of the state), 2 hours at Greenville, and 11 hours at Columbus. No freezing temperatures were recorded for any length of time from Jackson southward in the central part of the state. This late freeze event damaged much of the state's agriculture.

Before the freeze event, the wheat crop was forecast to be one of the best in recent history, with 350,000 acres planted and record yield potentials of 65–70 bushels/acre. Freeze injury occurred primarily north of a line spanning from Clarksdale to Columbus (Figure 7.14). Damage in the freeze area varied substantially, with yield reduction ranging from 10% to 100% depending upon the stage of development of wheat at the time of the freeze. Wheat yields in the freeze area ranged from 0 to 85 bushels/acre, but averaged about 45 bushels/ acre, so average wheat yield loss was about 40%. Some growers chose to abandon heavily damaged wheat and benefited from earlier planted soybeans in the same fields.

Mississippi growers planted about 980,000 acres in corn in 2007, nearly 95% of which had been planted at the time of the freeze. Injury was most significant in the same zone as for wheat (Figure 7.14). Because the growing point of the plant was protected below the soil surface at that point in the crop's

WINTER WEATHER SAFETY

In Mississippi, any amount of snow proves a challenge on the roadways because the state is not prepared to deal with the snow. Besides, in Mississippi, snow does not usually stick around for very long. For that reason, it is best to stay at home when it snows and not drive anywhere. The same is true of icy road conditions.

When winter storms are serious, especially when they bring freezing rain or sleet, the accumulation of ice on branches and power lines can cause power outages, sometimes for extended periods of time. For this reason, it is good to have a battery-operated radio, flashlight, and extra batteries. Other supplies that you might also place in a hurricane emergency kit, such as food, water, or medicines, are useful. An emergency heat source can help you stay comfortable inside if the power outage causes you to lose your heat source, but portable heaters can also be dangerous. Only burn in stoves and fireplaces designed for indoor use and be sure that they are properly ventilated. Also, only use the type of fuel that is meant for the appliance. Improper use of portable heaters can result in carbon monoxide buildup or fires.

In the event you are traveling and get caught in a car during a winter storm, it is best to remain in the vehicle. The National Oceanic and Atmospheric Administration and the American Red Cross recommend running your car for only 10 minutes at a time and making sure that your tail pipe is not blocked. They also recommend cracking the car window to prevent carbon monoxide poisoning and making the car visible to rescuers should you really get stuck.

Also, when temperatures get very cold, special accommodations should be made for pets. If the temperature is going to be below freezing, dogs that are kept outside will need their water replaced more often so it does not freeze. They should also be provided shelter if the temperature or wind chill is expected to drop below 32°F. In the event the temperature or wind chill is forecast to be below 0°F, outside pets should be brought inside or kept in a climate-controlled shelter. Also, certain breeds, young puppies, and some older dogs have a lower tolerance for cold weather and should not be kept outdoors in freezing temperatures. Some communities also have ordinances that that require pets to be brought inside when winter weather or heat advisories are issued. It is a good idea to be aware of these requirements.

development, most plants recovered from the freeze. About 15–20% of the corn acreage was replanted statewide due to the late freeze. Because replanting occurred early, the overall impact of the freeze on corn harvest was minimal, with no loss documented as a result of the freeze. (This assessment of the impacts of the freeze on wheat and corn was provided by Dr. Erick Larson, Grain Crops Agronomist with Mississippi State University.)

During an average year, the wheat yield potential in Mississippi is 22,750,000 bushels. The U.S. Department of Agriculture's National Agricultural Statistics Service (USDA/NASS) estimated a statewide yield of 18,480,000 in 2007, with some acreage not harvested. Based on a price of $5 per bushel, the estimated loss was $21.3 million. According to USDA/NASS, around 980,000 acres of corn was planted in Mississippi in 2007. If 20% needed to be replanted at an estimated cost of $40/acre, that would total around $7.7 million. Thus, the late freeze in 2007 caused total losses in the wheat and corn crops of about $29 million.

Damage to blueberries during the late freeze event was actually beneficial. Freeze damage thinned the bushes somewhat, allowing for larger berries on the remainder of the plants. No economic loss was reported. Tomatoes and watermelons, grown mainly in the southern part of the state, did not suffer freeze damage, but production and yield were slowed by the cold temperatures that did penetrate to the southern parts of the state during the freeze event. (This assessment was based on percentage damage estimates provided by Dr. David Nagel, Extension Professor, Plant and Soil Sciences, Mississippi State University.)

Further Reading

Both the Jackson NWS and the Memphis NWS have summaries of major winter storms (as well as other events) on their websites:
Memphis: http://www.srh.noaa.gov/meg/?n=events
Jackson: http://www.srh.noaa.gov/jan/?n=local_weather_events

PREHISTORIC HURRICANES IN THE GULF

From the 1500s to the early 1800s, the hurricane record was not so complete. That was because the Mississippi coast was not densely populated and much reliance was placed on ship records or the written record of significant damages to let us know when a hurricane struck land and how bad it was. Since better quality records have been kept, there have not been that many very intense hurricanes. In determining how common or rare they are, it would help to have a record that goes back further than 150 years. Scientists have begun to create this record using proxy methods to identify the occurrence of prehistoric hurricanes. The study of hurricanes through geological proxies or historical documents is called paleotempestology. Some of the proxies include sand, pollen, and a type of free-floating algae.

Kam-biu Liu is a geographer at Louisiana State University who has pioneered the field of paleotempestology. One central premise upon which his research has been built is that the presence of sand in cores taken from lake bottoms can detect the occurrence of past hurricanes. How this happens is simple. Under normal conditions, tiny particles settle at the bottom of a lake and become part of the lake floor. These particles are normally very fine. When a hurricane occurs, it pushes a large quantity of sand inland from coastal regions. Some of this sand is deposited in lakes located near the coast, where it also settles to the lake bottom. Over time, successive layers of coarser beach sand and fine lake mud are deposited at the bottom of the lake. Scientists can determine when the hurricane occurred by dating the sand layers in a core taken from the bottom of the lake. Coastal marshes can also be used in this way.

Dr. Liu's (2007) research along the Gulf Coast indicated 10 to 12 category 4 or 5 hurricanes had occurred over the past 3800 years. In the last 1000 years as well as the 1200 years between 3800 and 5000 years ago, there were not as many hurricanes. One of the specific sites Dr. Liu's team has used in their research is the Pascagoula Marsh at the base of the Pascagoula River in Jackson County, Mississippi. The results of study on this marsh show similar results to other samples along the Gulf Coast, with a very active period from 3720 to 1350 years ago. In the last 5000 years, 17 intense hurricanes were recorded in the Pascagoula Marsh (Hathorn 2008). This was based on sand layers as well as the presence of certain types of pollen and algae in the samples taken from the marsh bottom.

8. LONG-TERM CLIMATE CHANGE

Earth is about 93 million miles from its source of energy, the sun. Only about 2 billionths of the sun's energy output arrives in the vicinity of Earth, and about half of that amount is lost as it moves through the planet's atmosphere to the surface. Yet this small fraction of the sun's energy is sufficient to drive the processes that result in our planet's hospitable environment, which allows life in a diversity of forms.

The sun's energy strikes a planetary surface that is about 71% water and 29% land. Interactions with these dissimilar surfaces produce differential heating and cooling spatially. Furthermore, the planet rotates on an axis that is tilted 23.5° from perpendicular to a plane with the sun's center, and it also revolves in an oval orbit around the sun, causing an unequal distribution of energy across the surface. These inequalities create motion in the fluids of the planet, the atmospheric and oceanic circulations, which further redistribute energy and moisture. This system establishes a pattern of climatic zones that are collectively referred to as the planet's climate. Earth's climate is therefore a consequence of complex land, atmospheric, and oceanic interactions with energy, resulting in an average planetary temperature of about 59°F (Moran 2006).

Some processes alter Earth's energy flows and exchanges for short time spans. Such short-term climate changes may be on the order of 1 or 2 years to thousands of years in extent, producing small yet consequential environmental variations across some parts of the Earth's surface. By contrast, long-term climate changes may endure for millions of years and produce drastically different environmental conditions over all or large portions of the Earth. By far the majority of the Earth's known climate changes have occurred when there was no human influence on the planet.

Natural climate change has evidently occurred on many different scales of both time and space in Earth's 4.5 billion year history. Evidence shows that there have been warmer and colder intervals, changes in atmospheric composition, and even intervals when there were different land, ocean, and mountain distributions. Daily weather changes are good examples of natural changes occurring on time and space scales of a few days and miles. Phenomena such as El Niño/La Niña are examples of natural changes occurring over periods of seasons to years and affecting as much as a hemisphere in area. Changes in the Earth-sun relationships and even in the arrangements of continents and oceans on Earth are examples of natural changes occurring over time scales of hundreds of thousands to millions of years and affecting the entire planet.

This chapter concentrates on these natural causes of climate change. We provide a reconstruction of past climates to give perspective on the temporal and spatial scale of past climate changes. Mississippi will be placed into the context of these global climate changes.

RECONSTRUCTION OF PAST CLIMATES

When considering a reconstruction of Earth's climate record, it must be noted that many different types of data and evidence are used. Climate data recorded by instruments are available for about 150 years at most, and the distribution of observation sites and instrument types has been highly variable during that time. In Mississippi there are continuous records going back only to 1895. Beyond the instrumental record, use of historical data extends the climate record back to around 1000 years or so. For example, there are written documents describing the settlement of Greenland by the Vikings when it was more hospitable to human occupation, as well as the Medieval Warm Period of the Middle Ages and the Little Ice Age that persisted from about 1400 to 1800. Some early Mississippi records give details about the weather around Biloxi and Natchez in the late 1700s and an account of the great Natchez tornado of 1842.

Assessing climatic characteristics and climate change in prehistoric times requires use of proxy data. These proxies include evidence from continental drift, eustatic (worldwide) sea level change, landscape features of erosion and deposition, cores and isotope analyses from ice sheets and ocean floors, sediments from lakes and oceans, evaporites from caves, and characteristics of past flora and fauna as indicated by fossils, pollen, and tree rings.

CLIMATE CHANGE OVER GEOLOGICAL TIME

Geological time is a concept that allows visualization of the enormous span of time that covers the Earth's existence, 4.5 billion years. Of prime importance to the reconstruction of climate, it must be noted that virtually nothing is known about climate for fully seven-eighths of Earth's history. This division is marked on a geological time scale as the boundary between the Cambrian and the Precambrian at 542 million years before the present.

Figure 8.1 shows what is considered to be the history of climate on Earth during the past 542 million years, the Phanerozoic Eon, which includes all geological time divisions after the Precambrian. For much of this time, the land

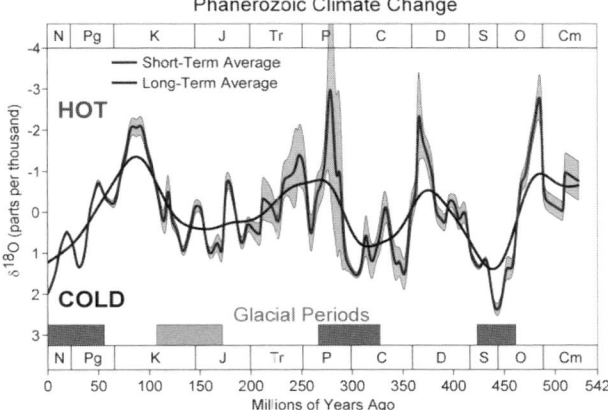

Fig. 8.1. Climate change over the past 500+ million years. ^{18}O is an isotope measured in sediment and ice cores that serves as a proxy for global temperature. The letters along the time axis indicate the period in geologic time (e.g., N, Neogene; Pg, Paleogene; K, Cretaceous). Image credit: Robert A. Rohde, Global Warming Art Project, http://upload.wikimedia.org/wikipedia/commons/9/9c/Phanerozoic_Climate_Change.png

that is now Mississippi was under deep ocean water, and even the North American continent had not been formed. The information on climate change at this temporal scale is very coarse at best, but it does indicate that there have been numerous times when Earth's temperature has been relatively warmer or colder than average. Figure 8.1 also shows that the planet's temperature at the present (from the geological perspective) is in a cooler phase.

Narrowing the time frame provides a higher resolution of past climate change information. The span of time from about 145 to 65 million years ago is known as the Cretaceous Period. At the start of the Cretaceous, Mississippi was still the sea floor. During this period, the supercontinent Pangaea had broken into the modern continents, but they were in different locations than today (Figure 8.2). This is also the time when dinosaurs inhabited the land and ocean. At the end of the Cretaceous, 65 million years ago, Mississippi was under a shallow sea between the present-day Appalachian and Rocky Mountains. By 23 million years ago most of Mississippi had emerged from beneath the water.

Figure 8.3 shows climate change over the last 65 million years, since the end of the Cretaceous Period. A notable feature of this graph is the difference in temperature during the Eocene Climatic Optimum, centered on about 50 million years ago, and during the last 10 million years. Again, this shows the potential variability in Earth's climate across large expanses of time.

Figure 8.4 shows the climate record of the past 5 million years by combining measurements from 57 globally distributed deep sea sediment cores (Lisiecki and Raymo 2005). The horizontal line at 0°C indicates modern global

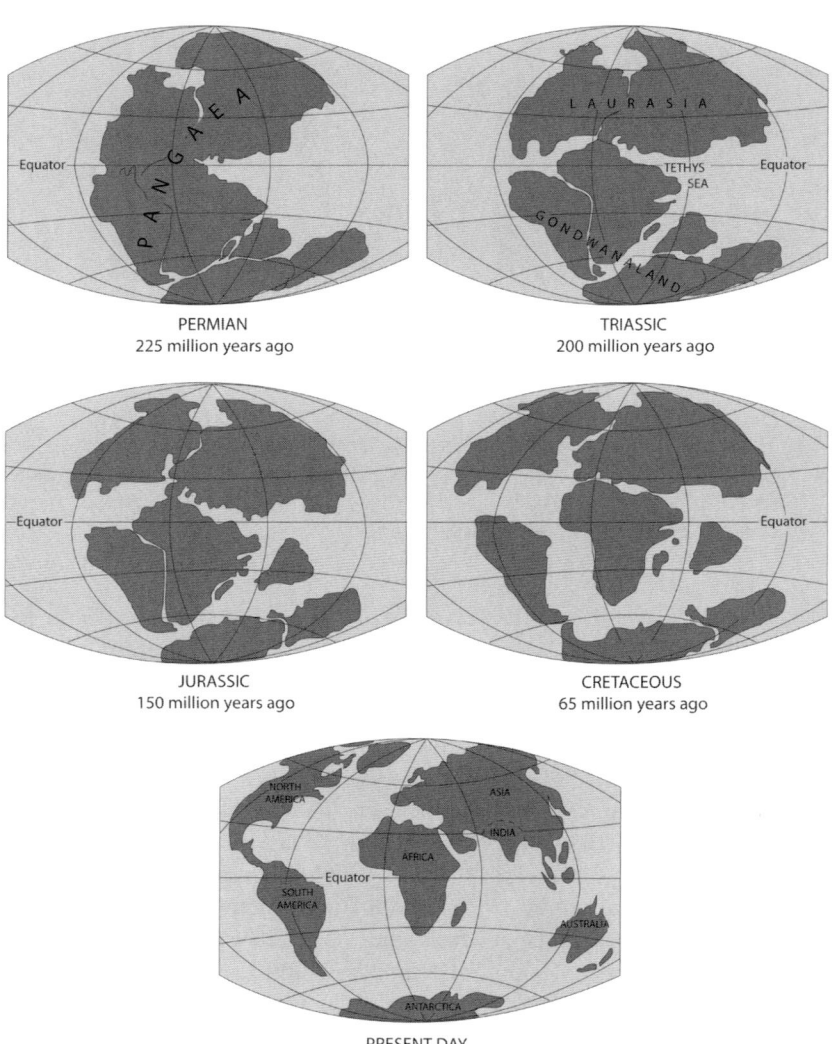

Fig. 8.2. The breakup of the supercontinent Pangaea and motion of the continents to their present-day positions. Image credit: U.S. Geological Survey, http://pubs.usgs.gov/gip/dynamic/historical.html

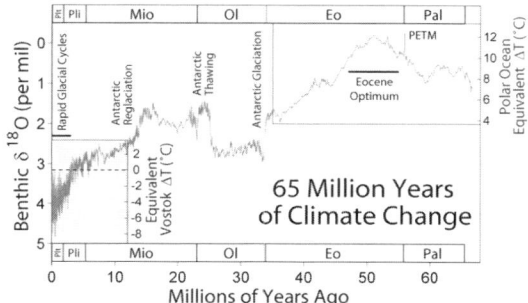

Fig. 8.3. Climate change over the past 65 million years. ¹⁸O is an isotope measured in sediment cores that serves as a proxy for global temperature. "kyr" means thousand years. The letters along the time axis indicate the epoch in geologic time (Pleistocene, Pliocene, Miocene, Oligocene, Eocene, Paleocene). Image credit: Robert A. Rohde, Global Warming Art Project, http://upload.wikimedia.org/wikipedia/commons/1/1b/65_Myr_Climate_Change.png

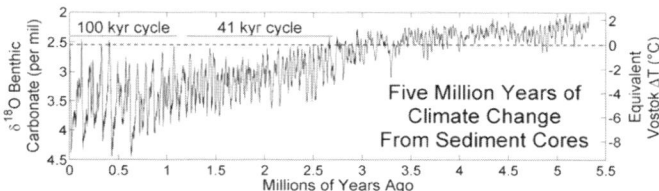

Fig. 8.4. Climate change over the past 5 million years. ¹⁸O is an isotope measured in sediment cores that serves as a proxy for global temperature. "kyr" means thousand years. Image credit: Robert A. Rohde, Global Warming Art Project, http://commons.wikimedia.org/wiki/File:Five_Myr_Climate_Change.png

temperature and shows a division of these past 5 million years into a warm and a cold phase.

Figure 8.5 shows a series of 100,000 year glacial cycles over the last 0.5 million years or so, indicating the most recent ice ages of Earth during what is known as the Pleistocene, which lasted from about 2.5 million until about 12,000 years ago. Earth experienced repeated colder glaciations and warmer interglacial periods during this time span, with ice sheets in North America coming as far south as 40°N latitude. Mississippi was never glaciated, so there are no glacial landform features in the state. However, when the ice sheets were advancing, sea level was as much as 400 feet below present levels and mean annual temperature was much colder than it is today. During interglacial periods, sea level and temperature were more like today.

One fairly unique geomorphic feature of Mississippi related to climate changes is the loess deposited on the bluffs along the eastern edge of the modern Mississippi River floodplain, known locally as the Loess Hills region (see Figure 2.5). Loess is a wind-deposited, calcareous silt laid down at the end of

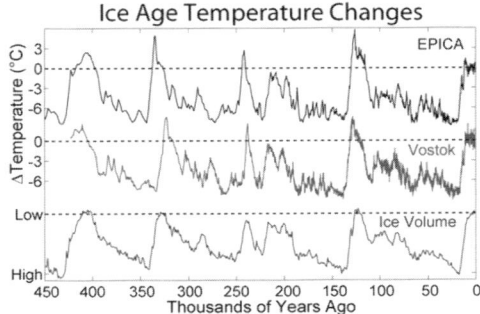

Fig. 8.5. Climate change over the past 450,000 years. The EPICA and Vostok lines indicate temperatures in the Arctic and Antarctic that were derived from ice core data. Image credit: Robert A. Rohde, Global Warming Art Project, http://upload.wikimedia.org/wikipedia/commons/f/f8/Ice_Age_Temperature.png

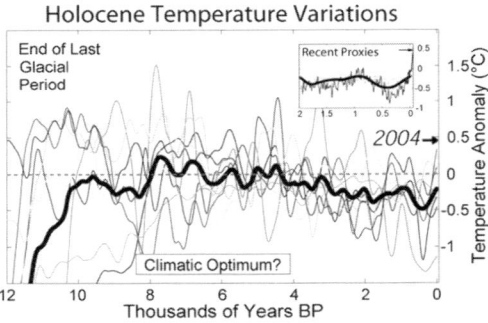

Fig. 8.6. Climate change over the past 12,000 years. The chart shows records of local temperature change in relation to the mid-20th century average (dashed 0°C line). The average temperature based on these is shown with the heavy dark line. Image credit: Robert A. Rohde, Global Warming Art Project, http://upload.wikimedia.org/wikipedia/commons/c/ca/Holocene_Temperature_Variations.png

the Pleistocene Epoch (Paulson 1974). Generally thought to have its origin in dust from finely ground rock, so-called glacial flour, the loess ranges in thickness from about 30 to 100 feet, averaging about 75 feet. Because of its calcareous content and its angular structure, the loess exhibits a tendency to resist erosion when standing vertically. The deposits are most noticeable in the vertical cliffs and road cuts near Vicksburg and Natchez. The loess mantle thins to the east over a distance of 20 to 40 miles. The material is about 75% quartz and about 20% carbonate in the form of dolomite. Interestingly, early farmers in the Natchez area were known to use the loess to neutralize acidic soil.

Figure 8.6 shows Earth's climate since the retreat of the last of the Pleistocene glaciations, in what is known as the Holocene or recent climate. By about

12,000 years ago global temperatures began rising and reached the Climatic Optimum between 7000 and 8000 years ago. During this time of warmer climate, civilizations emerged, agriculture began, and human population began to grow. The first Native Americans settled in what became Mississippi around 11,500 years ago, as the South was emerging from the last of the continental glaciations. During that time, the climate was cooler and the area was a mix of forest and grassland (Grice 2006). Another warm period known as the Medieval Warm Period occurred about 1000 years ago, and the dip in temperature known as the Little Ice Age is evident in Figure 8.6 at about 400–200 years ago.

In summary, it is generally believed that the Earth's climate experienced several warm and cold phases of immense time during the last 570 million years, culminating in the Eocene Climatic Optimum, centered at about 50 million years ago. At that time there was probably no ice on the planet, and since that time the climate has been cooler. Over the past 600,000 years Earth has experienced episodic ice ages interspersed with short warm periods, or interglacials. The last ice age ended 12,000–15,000 years ago, and Earth is currently in a warm interglacial. During this interglacial period, Earth experienced the Climatic Optimum about 7000 years ago, the Medieval Warm Period from about 1000 to 1300 A.D., and the Little Ice Age from about 1400 to 1800 A.D. Over the past two centuries, Earth has experienced warming since its emergence from the Little Ice Age.

These global climate changes and consequent sea level variations are reflected in the geology and fossil record in Mississippi. They are responsible for the archeological history, variety of soils, distribution of water and other natural resources, and landforms found in the state today.

INSTRUMENTAL RECORD OF CLIMATE CHANGE IN MISSISSIPPI

In recent years researchers have discovered several recurring seasonal to decadal length oscillations, or global-scale patterns of climatic variability, that can substantially affect local and regional weather patterns. These include the El Niño /La Niña phenomena, the Pacific North American pattern, the Madden-Julian Oscillation, the North Atlantic Oscillation, the Quasi-Biennial Oscillation, the Pacific Decadal Oscillation, and the Atlantic Multidecadal Oscillation. (See chapter 2 for a discussion of some of these climatic oscillations.) These oscillations, sometimes referred to as modes of climate variability or teleconnections, can shift weather patterns and disrupt local climate features (Gutzler et al. 1988; Hurrell 1996). Climatic variability in Mississippi is often related to these events.

For example, the positive phase of the Pacific North American pattern produces above-average surface temperatures over western North America

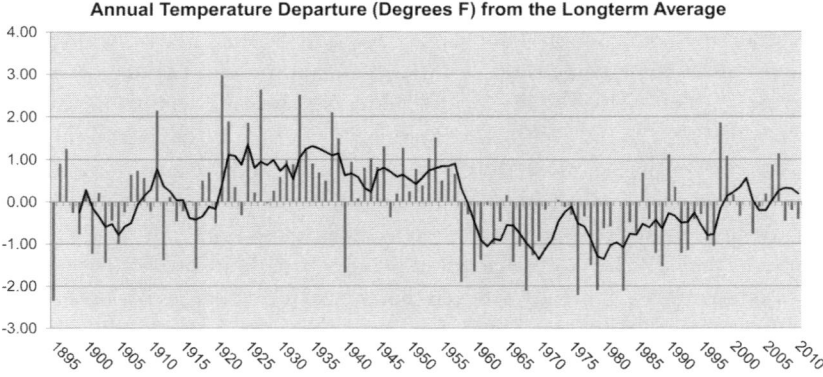

Fig. 8.7. Annual temperature variations in Mississippi, shown as the departure from the long-term average, 1895–2007. Data from National Oceanic and Atmospheric Administration, National Climatic Data Center, U.S. Historical Climatology Network

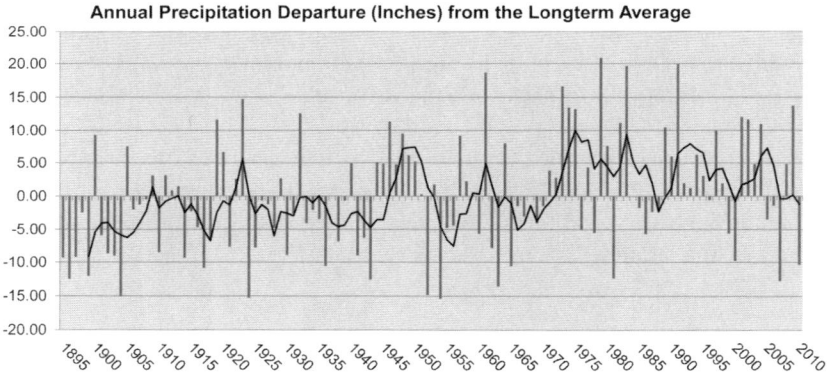

Fig. 8.8. Annual precipitation variations in Mississippi, shown as the departure from the long-term average, 1895–2010. Data from National Oceanic and Atmospheric Administration, National Climatic Data Center, U.S. Historical Climatology Network

in response to an unusually strong high pressure ridge over the region, while an upper-level trough centered over the southeastern United States produces below-normal temperatures over the Gulf Coast states. In contrast, the positive phase of the North Atlantic Oscillation pattern produces above-average temperatures over the eastern United States. These effects can last for an entire season or even persist for several years, producing variation in the climate of a region. The effects of these oscillations and patterns can be seen in the 1895–2007 temperature and precipitation records of Mississippi (Figures 8.7 and 8.8). For example, the period from the 1920s through the mid-1950s was warmer than average, whereas temperatures from the mid-1950s to mid-1990s were below average. Similarly, several wetter and drier periods are visible in Figure 8.8. Much of the first half of the 20th century was drier than average, while the 1940s and much of the latter half of the 20th century experienced wetter than average conditions.

FUTURE CLIMATE CHANGE

The average temperature in the Southeast did not change significantly in the last 100 years, according to the U.S. Global Change Research Program (USGCRP 2009) If anything, some parts of the Southeast have even cooled over the last 100 years (National Climate Data Center). As Figure 8.7 shows, the average annual temperature in Mississippi has fluctuated quite a bit over the last 100 years, but there has not been a significant temperature increase in Mississippi. Since the 1970s, the average annual temperature of the Southeast has increased by about 2°F (USGCRP 2009). The recent increase in annual precipitation shown in Figure 8.8 may be because Mississippi has seen average autumn and winter precipitation increase from 1901 to 2007 (USGCRP 2009). Mississippi has experienced no change in spring and summer precipitation. Unlike Mississippi, much of the Southeast has experienced drier conditions in spring, summer, and winter and wetter conditions only during autumn (USGCRP 2009).

Climate models predict warming across the Southeast through the 21st century (USGCRP 2009). This includes an increase in the number of 90°F days as well as a rise in overall temperatures. Coupled with projected global warming, these increases could have several consequences for the state, such as higher rates of illness due to heat stress, a decline in the diversity of aquatic species, and an increase in the effects of landfalling hurricanes if warming produces more intense storms (USGCRP 2009). Scientists are still debating the impact global warming might have on hurricanes, but the likeliest consequence is that hurricanes that form will be stronger (Elsner 2008). Stronger hurricanes could cause greater loss of barrier islands and greater coastal flooding.

While scientists are confident that on a global scale temperatures will increase during the rest of the 21st century, predictions of climate change are more uncertain in individual regions. Uncertainties exist in part because the predicted change is related to the various models' ability to predict changes in some of the features mentioned above, such as the North Atlantic Oscillation and El Niño/La Niña (IPCC 2007). Also, the more precise the prediction in terms of time or location, the more uncertainty is associated with the estimate. So, it is not possible to state with certainty what conditions will be like in Mississippi in the next 100 years.

Further Reading

National Climate Data Center. Global Warming: Frequently Asked Questions. http://www.ncdc
.noaa.gov/oa/climate/globalwarming.html

U.S. Global Change Research Program (USGCRP). 2009. Regional Climate Impacts: Southeast. In *Global Climate Change Impacts in the United States*. T. R. Karl, J. M. Melillo, and T. C. Peterson (eds.). New York: Cambridge University Press, pp. 111–116. http://www.globalchange.gov/im ages/cir/pdf/southeast.pdf

9. APPLIED CLIMATE

Climatology is the study of the average and extreme weather conditions that a place experiences over a given number of years, whereas applied climatology examines the influence of climate on that place. This could include where we live, how we dress, and what we do for work or fun. More broadly, applied climatology can help explain how weather and climate impact the economy or the environment. One of the most important influences climate has in Mississippi is on agriculture, from the backyard garden to the large-scale farm.

THE MISSISSIPPI BACKYARD GARDENER

Mississippi generally has a long gardening season. The ground does not freeze for long periods of time, if at all, so some garden plants can be grown all year long. During winter, most of Mississippi will see the overnight temperature fall below 32°F. How often this occurs will depend on what part of the state you live in. Some winters, the temperature never falls below 32°F in southern Mississippi. Most winters in northern Mississippi have several streaks of below-freezing temperatures. Figures 9.1 and 9.2 show the median date of the last spring and first autumn freeze. The median date is at the exact middle of all the reported dates. So, if the median date when the temperature falls below 32°F is November 1, this means that half of the time, the first autumn freeze is before November 1 and half of the time it is later. Although the maps show a range of freeze dates for each zone, this does not mean that in any particular year the actual dates of the first or last freeze will be in those ranges.

The average first and last date of freezing temperatures varies from north to south but also has a northeast-to-southwest pattern in the state. More than a month separates the median date of the last spring freeze as you travel across the state. The latest median spring freeze date is April 11. This distinction is shared by the Hickory Flat and University of Mississippi weather stations in north-central Mississippi. At the other end of the state, the Gulfport Naval Center reports a median last spring freeze date of February 21. In autumn, the University of Mississippi station is also the location with the earliest median first autumn freeze, on October 20. Biloxi has the latest date on December 10. Figure

Legend

Median Date of Last Spring Freeze

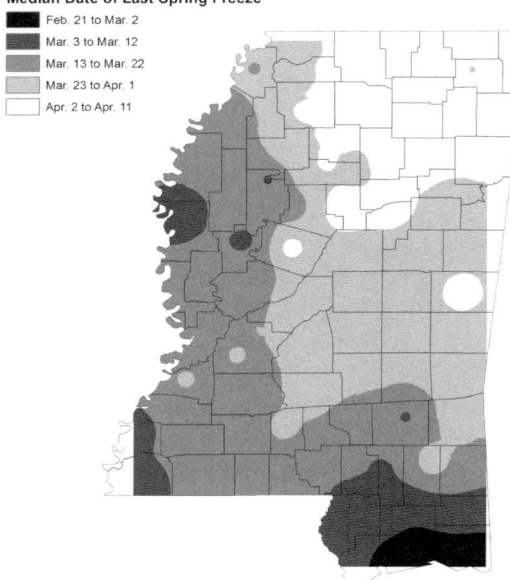

Feb. 21 to Mar. 2
Mar. 3 to Mar. 12
Mar. 13 to Mar. 22
Mar. 23 to Apr. 1
Apr. 2 to Apr. 11

Fig. 9.1. Median date of the last spring freeze. Data from National Climatic Data Center

Legend

Median Date of First Fall Freeze

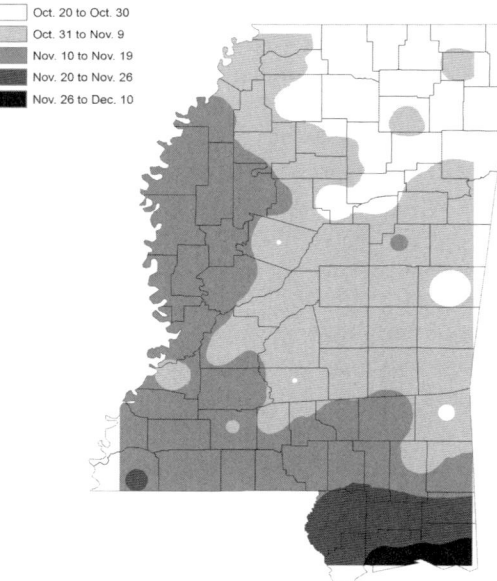

Oct. 20 to Oct. 30
Oct. 31 to Nov. 9
Nov. 10 to Nov. 19
Nov. 20 to Nov. 26
Nov. 26 to Dec. 10

Fig. 9.2. Median date of the first autumn fall freeze. Note that in this figure the shading goes from earlier to later dates to show the similarity of regions in Figure 9.1. Data from National Climatic Data Center

Legend

Median Number of Frost-Free Days

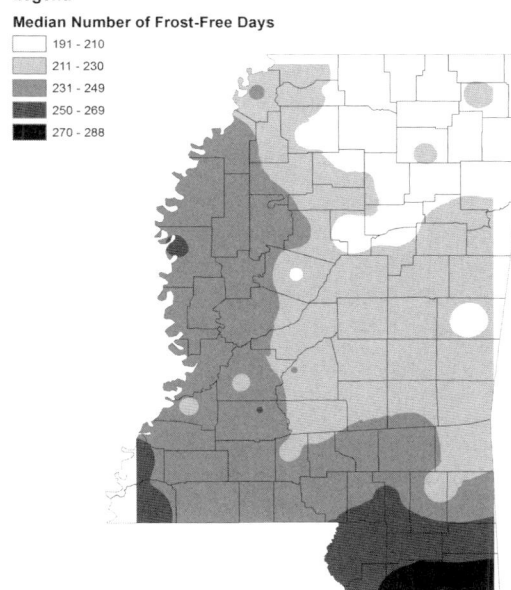

Fig. 9.3. Median number of frost-free days.
Data from National Climatic Data Center

9.3 shows the median number of frost-free days. The pattern is quite similar to those shown in Figures 9.1 and 9.2, and the frost-free pattern corresponds with the state's growing zones.

There is a difference between a frost and a killing frost. A frost occurs when water vapor in the air turns into ice crystals on plant surfaces. If the temperature is only briefly below freezing, leaves may show damage, but plants may still survive. During a killing frost, temperatures are below freezing long enough for ice to form within plant cells, thus killing the plants. A killing frost occurs when temperatures dip below freezing for a longer period of time and can occur without icy white frost as evidence. During the state's transitional seasons of autumn and spring, the weather patterns change quickly. A cold front will pass through and high pressure will build in behind it. Early in the season, the below-freezing temperatures usually only occur when high pressure is located very close to the area and on nights when winds are calm and skies are clear. These are the best conditions for a frost to occur. During this transitional time of year, given enough sunshine afternoon temperatures can quickly rebound to temperatures high enough to promote plant growth.

Plants can tolerate temperatures close to or even below freezing, but they have a certain temperature that they require to grow and to produce flowers, fruit, or vegetables. Warm season plants such as tomatoes, corn, and green

FREEZE WARNINGS

The National Weather Service (NWS) issues warnings when the weather will be below freezing, but only under certain conditions. A freeze occurs when the surface temperature will drop below 32°F for a significant period of time. What is considered significant varies from one forecast office to another. For example, the Jackson NWS forecast office issues a freeze warning when the temperature is expected to be below freezing for 1–2 hours or more. The NWS usually stops issuing freeze warnings once an area has had a killing freeze. They begin to issue the warnings again in the spring once plants have started to grow.

A hard freeze warning is issued when subfreezing temperatures are expected for a long duration. The Mobile NWS forecast office, which forecasts for five counties in southeastern Mississippi, issues a hard freeze warning when at least 5 hours of temperatures below 26°F are forecast. The Memphis NWS forecast office, which forecasts for the northern third of Mississippi, uses a 28°F threshold. During a hard freeze warning, plants—as well as people, pets, and pipes—need protection from the cold.

beans require the average daily temperature to be at least 50°F. Very warm season plants such as okra and peppers require temperatures above 60°F, whereas cool season plants such as cabbage and peas have a base temperature of only 40°F. In order to grow, plants also require temperatures to be warm enough over a sufficient period of time, that is, plants have particular growing degree day requirements. A growing degree day occurs when the average temperature is greater than the base temperature required for growth. For instance, if the base temperature for beans is 50°F, a growing degree day will occur once the average temperature [(maximum + minimum temperature)/2] is greater than 50°F. If the high temperature is 78°F and the low is 44°F, the average temperature for that day is 61°F. When the base temperature is subtracted, the day provided 11 growing degree days for that vegetable. Growing degree days are accumulated over the season and are related to the plant's developmental stages. For example, cotton first blooms after it has accumulated 950 growing degrees days and can be harvested once it has reached about 2600 (both based on a base temperature of 60°F) (Ritchie et al. 2007).

Some plants also have an upper temperature limit, and once it is exceeded, they stop growing or making fruit. For instance, tomatoes do not grow as well when the temperature goes above 90°F, and above 95°F they may stop producing fruit. Some plants also need a certain amount of cold weather. Anyone who has tried to grow lilacs that failed to bloom can take comfort in the knowledge that it was not the gardener, it was the fact that lilacs need a certain number of winter cold hours before they will produce flowers the following spring. Hardiness zones 7 and higher (Cathey and Jordan 2001) are generally too warm in the winter to provide enough cold hours for lilacs. Some tulips also require a cold period that is too long to be achieved in Mississippi winters.

The U.S. Department of Agriculture published a list of "Basic Plant Requirements" for a broad array of plants that one needs to take into consideration before planting (Cathey and Jordan 2001). Most of the basic requirements are weather related and include day length, radiation, temperature, heat, rainfall, and pH.

Day length, the period of time from sunrise to sunset, is very important to the life cycle of the plant. It can promote growth, as well as let the plant know that the time has come to go dormant. The amount of daylight in Mississippi varies by about 4 hours from the shortest day in winter to the longest day in summer.

The amount of solar radiation a plant is exposed to will determine whether it can flower and set fruit or if it will not do well where it is planted. For example, vegetables require sunny conditions and will not thrive in shady locations. Tomato plants must have at least 6 hours of unfiltered sunlight to successfully set fruit. This is true of most summer fruits and vegetables. Many plants also prefer morning sun. Cloudy or rainy days reduce the amount of sunlight the plant receives and can negatively affect its development. Plants also vary in the intensity of heat they can tolerate. To help gardeners with these requirements, plants sold at garden centers and nurseries are typically labeled full sun, partial sun, or shade. Plants that need full sun in a cooler climate may need partial sun in Mississippi. Hydrangeas are an example of a plant that requires more direct sunlight the further north it is planted.

Plants do best when temperature and rainfall conditions are optimal. Like temperature, precipitation requirements also vary considerably among plants. Some plants will thrive in Mississippi without extra watering in most years. Lantana is a good example of a flowering plant that does well in the heat without a lot of watering. Many trees and shrubs do fine without watering once they are established and the roots can penetrate the soil deep enough to reach water. In most years, the rainfall in Mississippi is not ideal for tomatoes. Irregular precipitation can lead to blossom end rot or cracks on the skin of the tomato. Warm season lawn grasses typically survive without watering in most summers; however, in drier periods, a deep soaking every 2–3 weeks may be necessary. It is better for most types of lawn grass to water for a longer period of time less frequently than to water multiple times per week.

The pH of a substance is the measure of how acidic or alkaline (basic) it is. At one extreme, battery acid has a pH of 1. At the other end, drain cleaner has a pH of 14. Acid rain has a pH of 4.2 to 4.4, and the pH of soil ranges from about 5 to 8. Most plants prefer pH that is slightly acidic (about 6.5). If the pH is not right for the plant, its roots will not be able to absorb the nutrients and water it needs. Plants grown in Mississippi that prefer acidic soil include blueberries, watermelons, and azaleas. Plants that prefer a more alkaline soil include cabbages, geraniums, and thyme. Areas with greater rainfall typically have more acidic soil, including most of Mississippi. Some regions, such as the Delta and the Black Prairie (see Figure 2.5), have more alkaline soil. Alkaline soils are usually associated with drier climates, but the limestone in the Black Prairie soils makes it more basic. Where soils are too acidic, agricultural or dolomitic limestone can be added.

Plants that will flourish in Mississippi are limited by yearly variation in temperature and moisture. Mississippi is located within hardiness zones 7 and 8, so that many plants can survive the winter here that would not in the Midwest or Northeast. These include camellia and jasmine. Gardenias survive most winters, whereas they would need to be kept indoors in more northerly climes. Many tropical or subtropical plants, such as banana trees or palms, are used in landscaping in the state, but they will not survive the occasional winter that is colder than normal. On the other hand, some plants do not do so well in Mississippi because the summers get too hot. Hostas grown in Mississippi require more shady conditions than in other parts of the country, and some annuals, such as pansies and snapdragons, are planted during the cool season and die off once it starts getting too warm in May or June. These are considered summer plants in states not much further north. Some perennials sold in garden centers around the state, such as delphinium and foxglove, are more suited to locations with cooler summers. Rosemary and lavender, two Mediterranean plants, have no problem with Mississippi's summer heat, but they prefer nights that are cooler and drier than are found in the state. They often need a little extra care to be sure their roots have proper drainage. Thus, the South's humidity can be a problem for some plants.

CLIMATE AND AGRICULTURE

On the one hand, Mississippi's climate is characterized as mild with rain all year and a long hot summer. On the other hand, the state's climate is highly variable with lengthy periods of drought or heavy rainfall. Regardless, the state's economy is strongly tied to agriculture and forestry. According to the Mississippi Forestry Association, 65% of the state's land is forested. While the majority is not used for agriculture, timber is the most valuable agricultural crop for over half of Mississippi counties. After forestland, cultivated cropland accounts for the greatest amount of land (Johnson et al. 2002). One of the regions that typifies large-scale agriculture is the Yazoo Basin, also known as the Delta. This area in northwestern Mississippi is the floodplain of the Mississippi River, with mostly flat terrain and alluvial soils. The Delta has become a well-known and dominant agricultural region in the United States. About 4 million acres in size,

Fig. 9.4. Broad view of a catfish farm in the Mississippi Delta. Image credit: Stoneville Experiment Station, U.S. Department of Agriculture

this region produces 70% of the cotton, 70% of the soybeans, 60% of the corn, 95% of the catfish, and 100% of the rice grown in the state.

To mitigate the variability in precipitation distribution and help guarantee more even production each year, many of the producers in the Delta are turning to irrigation. About 45% of crop acreage in the Delta is under irrigation from the shallow alluvial aquifer. Almost 85% of all permitted wells in the state are located in the Delta, with hundreds of new irrigation wells being permitted each year. Climatological studies have shown that growing season rainfall in the Delta region delivers enough water for the crops in many of the years, but use of groundwater for irrigation overcomes the problem of distribution of the rainfall during the growing season and in different years. Recharge from rainfall cannot keep up with the withdrawal pressure on the aquifer, and the volume in the aquifer has been dropping an estimated 300,000 acre-feet per year since the mid-1990s. Efforts are underway to identify conservation methods using the natural climatic attributes that will allow sustainable use of the aquifer for continued and increased agricultural production into the future.

CLIMATE AND THE CATFISH POND

One agricultural product for which the Mississippi Delta has become nationally and internationally known is farm-raised catfish. The geographic and climatic attributes of the region combine to make aquaculture a very productive endeavor, and over the last 40 years many acres of catfish ponds have been constructed. The industry began in 1965 with one 20-acre pond, and there were more than 100,000 acres of catfish ponds in the Delta at the height of the industry in the late 1990s. These ponds became a large and noticeable user of the groundwater from the shallow alluvial aquifer, and during the drought of 1988 there was a threat of shutting down some of the wells supplying groundwater to

Fig. 9.5. Catfish production specialist Charlie Hogue tests water in a Noxubee County Catfish pond. Temperature and precipitation can affect the fishes' health as well as their appetites. Photo credit: Mississippi State University Office of Agricultural Communications

the ponds. Climatological studies have shown that the climate delivers enough rainfall, if captured, to supply most of the water needed for aquaculture in the region.

The catfish industry in the Delta has become the vanguard for groundwater conservation in aquaculture through use of a climatological scheme adopted by most producers. The strategy is to refrain from keeping the ponds full, allowing room for storage of rainfall when it occurs. The scheme allows water levels in the ponds to fall 6 inches before any groundwater is pumped in, then only raising the pond level by 3 inches, and again leaving room to capture rainfall. Use of this drop/add management strategy has been shown to save between 50% and 75% of the groundwater previously used in catfish production. This is an example of Mississippi agricultural producers using the climate to their benefit.

SOUTHERN ARCHITECTURE

Before Europeans settled in Mississippi, the Chickasaw lived in northern Mississippi. Well adapted to the Southern climate, the Chickasaw had different ~nes for summer and winter. Their summer homes were rectangular and ᵇᵉ woven walls and roofs made of thatch. The winter homes were

Fig. 9.6. The Stietenroth House in Natchez, Mississippi, is an example of a house in the Creole style. Although built in the second half of the 19th century, the house does not have the ornamentation that was common in Victorian-era homes. Photo credit: Library of Congress, Prints and Photographs Division, reproduction number HABS MS-271-9

partially belowground and circular, with walls made of grass plaster. The Chickasaw designed their homes to stay cooler in summer and warmer in winter.

One of the first Western influences on architecture in Mississippi was that of the early French settlers, whose houses followed French Colonial building traditions. The styles were also adapted to the conditions they encountered along the Gulf Coast and up the Mississippi River. To help cope with the heat, houses were built with ventilation in mind. They included high ceilings, double doors, and large casement windows (Foster 2004). Houses were raised where necessary to protect from flooding as well as to improve air circulation around the house, and they had large porches covered by an extended roofline to take advantage of the shade and outdoor air. Some houses, especially in the French Creole style, were built with hipped roofs extending outward away from the building on all four sides, providing even more shade.

Porches were not typical of European homes, but this architectural feature spread throughout the United States beginning with the South. The homes that first made the porch popular were those built by French colonists and their descendants near the Gulf. Early homes were smaller and single storied, but plantation homes were often larger. The Creole Cottage, Creole Plantation House, and Cajun House were examples of French Colonial architecture built in Mississippi in the 18th and early 19th centuries (Figure 9.6).

Fig. 9.7. The Davis Dog-Trot Style House on display at the Oren Dunn Museum in Tupelo was built northeast of Houston, Mississippi, in 1870 and later moved to the museum grounds. Note the breezeway at the center of the property. Photo credit: K. Sherman-Morris

The Creole Cottage style remained popular after other settlers moved into the area because it had worked so well in the Southern climate (Sanders, n.d.). New groups of people brought their own particular styles, although many houses retained the overhanging roof and porches. Eventually, as other house styles became popular in the South, especially among the upper class, some of the efforts to achieve cross-ventilation, such as designing interior rooms to allow the cross flow of air, were abandoned.

As people settled the frontier of the Appalachians and upland South, log cabins became popular due to the abundance of forests. One log cabin style well-suited for Mississippi's climate was the dog-trot style, which has an open area breezeway in the middle of two square rooms to allow for air circulation (Figure 9.7). However, the original reason for having two separate buildings was not really climatological. The dog-trot was usually built one one-room cabin at a time. When the second cabin was built, it was usually left detached because joining it to the existing cabin would be difficult. The dog-trot house remained popular through the 19th century and was modified to include wood rather than logs and ornamental details such as porch columns to adapt to changing styles and levels of affluence.

The shotgun style house is another Southern architectural tradition that was built to take advantage of climate (Figure 9.8). These long one-room-wide

Fig. 9.8. Elvis Presley's boyhood home in Tupelo is a famous example of a shotgun-style house. Photo credit: K. Sherman-Morris

houses have doors that are aligned with each other and windows across the narrow frame to allow for effective cross-ventilation (Foster 2004). These houses are also raised, which allows the air to circulate under the house as well. Inexpensive to build, shotgun style houses can be found in both rural and urban areas throughout the state.

Before air conditioning, houses built for the South took advantage of shade and cross-ventilation. Some of the historic adaptations to the climate can be used in building homes today. Location on the lot is a starting place. A publication by Southern Climatic Housing at Mississippi State University (Lewis and Kitchens 2006) suggests that to maximize cross-ventilation, houses should be built twice as long as they are wide, with the longest dimension oriented east to west. Building in this orientation can also allow for low-angle sunlight to help with home heating costs in the winter. This simple strategy can save 30% on energy costs. Other ways to block the summer sun, such as planting shade trees and building overhangs or trellises, can also help to keep houses cool in the summer.

Fig. 9.9. The Riverwalk (top) and The Rainbow (bottom) Casinos in Vicksburg used massive amounts of sand to prevent flooding during the Mississippi River flood of 2011. Photo credit: K. Sherman–Morris

WEATHER AND TOURISM

Mississippi's mild winter weather offers a long season for golfing and fishing. Visit Mississippi, a promoter of tourism statewide, emphasizes on their website that the temperate climate offers "near perfect golf weather year round." Similarly, the mild climate and Mississippi's many streams provide a year-round fishing season as well (Kinton 2002). The mild climate may also allow some of the freshwater fish to grow larger than in cooler climes because of the extended growing season. This is also beneficial for farm-raised catfish.

The largest draw for visitors to the Mississippi coast is casino gambling, which provides nearly $300 million in tax revenue and admits more than 30 million people a year, according to statistics kept by the state gaming commission and the Mississippi Casino Operators Association. In a typical year, casino gambling is not affected much by weather. Studies have shown that people who visit the state to gamble do not participate in a lot of outdoor activities, including visiting the beaches (Von Herrman et al. 2000).

While not much impacted by daily changes in the weather, hurricanes can prove disastrous for casinos located along the Gulf Coast. Even a weaker storm like Hurricane Georges can have a big impact on state coffers. When Hurricane Georges made landfall near Biloxi and Ocean Springs in 1998, casinos paid workers' hourly salaries while they were shut down for a few days. According to Von Herrman et al. (2000) the loss of tips cost the state money in income tax collections. This report estimated that for the year 2000 the loss to state and local governments would be over $400,000 for each day the casinos had to close in a hurricane. Of course, a much more powerful hurricane did impact the Gulf Coast in 2005, Hurricane Katrina. A report by the Gulf Coast Business Council in 2008 stated that 80% of the available hotels were either destroyed by or otherwise closed after Hurricane Katrina. Three years later, the hotels rebounded to about 70% of their pre-Katrina levels. The entire casino industry was put out of commission for 3 months. A hurricane is not the only weather-related disaster to impact the casino industry. Mississippi River flooding in 2011 caused all but two of the state's riverfront casinos to close.

Hurricanes can also have a positive outcome on tourism. After Hurricane Katrina caused substantial damage to roads, beaches, and property, Harrison County approved a plan to beautify and manage redevelopment along 26 miles of sand beach for 20 years (Gulf Coast Business Council Research Foundation 2008). The expansion and renovations made to the Prime Outlets in Gulfport after the shopping center lost a majority of its roof in Hurricane Katrina contributed to an increase in traffic and an award for contribution to the Gulf Coast economy (Gulf Coast Business Council Research Foundation 2008).

Weather also has an impact on determining oil and gas prices. Changes in the supply or demand for oil can nudge the price up or down. A harsh winter

AUTUMN COLORS

Although Mississippi does not have much of a reputation for autumn foliage, areas such as the Natchez Trace Parkway, Tishomingo State Park in the Appalachian foothills, and Wall Doxey State Park near Holly Springs are all favored spots for viewing the autumn colors.

Trees do not reach their peak color in Mississippi until mid-November. The exact timing depends on the weather for that autumn. Change in the amount of daylight (described below) plays the greatest role in signaling to leaves that it is time to change color, but the amount of rainfall and temperature can determine whether the season is good or bad for autumn foliage and whether leaves change early or late. The best autumn conditions for leaves to change color are when days are sunny and nights are cold. When summer is very dry, as in 2007 or 2010, leaves can turn brown and drop without displaying the reds and yellows that are pretty to view. Likewise, an ample supply of rainfall during the tree's growing months in spring and summer is also important for lovely autumn colors. Freezing temperatures can also make the leaves fall earlier, but not necessarily turn red, orange, or yellow first.

The colors of leaves vary according to the type of tree and the main type of pigment in the leaves. During spring and summer, chlorophyll, the substance that helps leaves obtain their energy from sunlight, makes the leaves appear green. In autumn, the decrease in intensity of sunlight and the shorter day-lengths reduce the plant's energy requirements and chlorophyll decreases. Other pigments take over to give the leaves their different colors. Leaves can even change at different rates on the same tree when one part of the tree gets more sunlight than the other.

Some native trees that provide autumn color in Mississippi include sweet gum, dogwood, and red maple. Sweet gum leaves blend colors from burgundy to yellow, while dogwood and red maple leaves turn red with hints of gold or burgundy. Cyprus trees, which may give the appearance of an evergreen during summer, also turn an orangey-brown color in autumn. The Bradford pear, a tree commonly used in landscaping, turns a variety of colors. Gingko trees, which are also planted for their stunning show in the autumn, turn a fairly consistent yellow.

Fig. 9.10. Autumn leaves.
Photo credit: K. Sherman-Morris

or even the forecast of a harsh winter can cause prices to rise, while a forecast of a mild winter can cause prices to drop. A dramatic impact can be felt when a hurricane threatens, even if it does not make landfall in the state. Hurricanes often place a hold on drilling operations in the Gulf of Mexico, which can cause prices to rise for people anywhere in the country. A hurricane can also mean lost wages for individuals affected by the shutdown. According to a 2006 report by the Federal Trade Commission, Hurricane Katrina resulted in the shutdown of over 95% of crude oil production in the Gulf, as well as a refinery in Pascagoula, Mississippi.

Further Reading

Cathey, H. M., and R. Jordan. 2001. *USDA Plant Hardiness Zone Map*. USDA Miscellaneous Publication No. 1475. Washington, D.C.: U.S. National Arboretum Agricultural Research Service, U.S. Department of Agriculture. http://www.usna.usda.gov/Hardzone/hrdzon2.html, accessed October 8, 2010.

Foster, G. 2004. *American Houses: A Field Guide to the Architecture of the Home*. New York: Houghton Mifflin.

Fig. 10.1. (clockwise from left) Barbie Bassett, WLBT meteorologist; John Dolusic, WTVA meteorologist; and Jason Dunning, former WCBI meteorologist. Photos courtesy of Barbie Bassett, K. Sherman-Morris, and Jason Dunning

10. WEATHER INFORMATION
From Observation to Forecast to You

This book has shared a lot of information about weather and climate in the state of Mississippi. This last chapter provides the back-story. We look at how weather forecasts are made, how they are shared with the people who want them, and how weather was predicted before there were instruments to record the state of the atmosphere.

TELEVISION WEATHER FORECASTS

The most common source of weather information for people is the television weather forecast, and the top source is local television. The weathercaster is the main face for weather information for many people. We watch them so regularly, sometimes it feels like we know them personally. Most people reading this probably have a choice in what local television news and weather to watch, but you probably also have a favorite that you watch most of the time. There are six local television markets based in Mississippi. This does not include the large area in northwestern Mississippi that gets its local news from Memphis or the three other out-of-state markets that provide the news in southern Mississippi (New Orleans, Mobile, and Baton Rouge). Each of the market areas has at least one station that airs a local news program. Many of the larger markets have three or more local news stations.

The largest television market within the state is the Jackson market, which has nine television stations and serves more than 300,000 households. Although Jackson has a population of only about 177,000, the Jackson market area stretches to the Louisiana border at Adams and Pike Counties and to Attala County in central Mississippi. The weather forecast does not usually vary much from one end of a television market to another, but a large viewing area can make following severe weather more difficult for the television meteorologist.

The oldest broadcast station in the state is WJTV in Jackson. Its first air date was January 20, 1953, with the inauguration of President Dwight Eisenhower its first program. Several other Mississippi television stations also made their debut in 1953. WLBT in Jackson made its first broadcast on December 19, 1953. On air to give the weather forecast that day was Woodie Assaf, the longest tenured weathercaster in Mississippi. Assaf was born in McComb, Mississippi, and got his start in radio. After a time at WQBC in Vicksburg, he began his career in Jackson working at WJDX radio. He worked at WLBT from 1953 until he retired in 2001, making him the weathercaster with the longest career at the same station anywhere in the country. Woodie Assaf died in 2009, but is remembered fondly by fans and former coworkers. "You watched because you loved to hear what Woody was going to say and what Woody was going to do," said current WLBT chief meteorologist Barbie Bassett. "You liked Woody . . . and that made you a loyal WLBT viewer."

A few other local news programs have weathercasters who have been on the air for quite a while. Dick Rice, chief meteorologist emeritus for WTVA in Tupelo, has the distinction of being the current weathercaster who has been on air longer than anyone else in the state. According to the WTVA website, Rice came to work for the station in 1979 to do weather on the *Mornin'* and *Noon* shows. Born in western Massachusetts, Rice has spent time in some exotic locations. Working as a meteorologist for the Navy for 21 years, Rice spent time on ships, overseas, and in facilities here in the United States. He even spent some time at the Joint Typhoon Warning Center on Guam. Rice stopped doing daily weather broadcasts in April 2011, but he continues to take part in school visits and fills in as needed.

The transition from a military forecaster to a television weathercaster was a common occurrence in the early days of television. It was a military meteorologist, Vilhelm Bjerknes, who first discovered that boundaries separate warm and cold air masses during World War I, identifying these boundaries as "fronts." Hardly a newscast goes by from the autumn through the spring when the weathercaster does not mention a warm front, cold front, or the lack of any fronts to bring us weather. Following World War II, several local stations hired military-trained forecasters. Bryan Owings, the former chief meteorologist for WCBI in Columbus, went on air after a full career for the Air Force. Owings joined WCBI the same year Rice started at WTVA and retired in 2004.

Another Navy meteorologist turned weathercaster is Mike Reader, chief meteorologist at WLOX in Biloxi. Reader came to WLOX as the weekend weathercaster in 1983. Before that, he forecasted weather conditions around the world and flew into hurricanes with the Hurricane Hunters. He also worked through Hurricanes Elena, Georges, and Katrina. Tommy Richards also has been at WLOX for quite some time. Unlike Reader, Richards got his start in radio. While at WLOX, he has filled a number of the stations positions, but as noted in his station biography, "His heart lies in reporting the weather." Several other weathercasters have also been on the air in Mississippi for more than 20 years, including John Dolusic at WTVA and David Hartman at WAPT. Such longevity helps explain why these individuals are trusted sources of weather information for so many viewers.

Three television meteorologists from different parts of the state and with different experiences, Barbie Bassett, John Dolusic, and Jason Dunning (Figure 10.1), recently discussed their careers with one of the authors. It was interesting to learn that they all shared some experiences, while each has his or her own unique take on television weathercasting. Bassett, the chief meteorologist for WLBT in Jackson, was raised on a farm in Marks, Mississippi, which taught her what a powerful influence the weather can have on a family's livelihood. In a year with bad weather, which could mean too much rain or not enough rain, she would watch her father worry over crops and came to realize that the family might not be getting many new things that year. Her experiences growing up on the farm were what drew Bassett to weather. On the other hand, John Dolusic, meteorologist at WTVA in Tupelo, got interested in weather because of an experience with a tornado when he was 3 years old. He reports being "weather crazy" as a child. Dolusic also remembered the influence of a favorite weathercaster, Dan Henry, whom he watched while growing up in Kansas City. Dolusic said he always loved weather and overcame problems with stuttering to get into television weathercasting. Jason Dunning was a morning meteorologist at WCBI in Columbus from 2009 to 2011 and before that a weekend meteorologist at WTOK in Meridian. (In April 2011, he left WCBI for ABC-7 in Ft. Myers, Florida.) He was also influenced by weather as a child. Dunning, who grew up in Florida, the thunderstorm capital of the United States, was first afraid of thunderstorms, but soon his fear grew into a passion for weather as he learned more about it.

Bassett, Dolusic, and Dunning all start their days by reviewing the current conditions and the computer models and thinking about their forecast. As the morning meteorologist, Dunning's day usually began quite early. On days when no bad weather was expected, he got to work around 4:00 a.m. to finish making his forecast and preparing for the first show at 5:00 a.m. Dunning also made the occasional school visit. Bassett, Dolusic, and Dunning also all create their own graphics that you see on television. Dolusic likened building his forecast graphics to playing a computer game. Bassett said she tries to think about how to make her graphics interesting and new for each newscast. She also stays busy by recording forecasts for the station's weather phone service, which has over 1 million callers per month, the 24-hour weather station, NBC Weather Plus, as well as forecasts for 24 newspapers around the state. Updating the website and social media is also a relatively new component of the weathercaster's day.

Parts of their jobs these weathercasters find challenging are unique to television, such as making things look new to the viewers each time, while other aspects are the same as in other careers, such as balancing a work and home life. One understandably challenging aspect is getting the forecast correct on a

Fig. 10.2. Inside the studio at WCBI. Jason Dunning is seated at the Weather Center in the background. Photo credit: K. Sherman-Morris

complicated day and communicating this complicated information is a way everyone will understand. Severe weather can also be challenging for the weathercasters, both personally and professionally.

Dolusic recalled the 1998 Ingomar tornado and the 2001 Pontotoc tornado as being "really intense." He remembered people calling the station in fear because they did not really have a safe place to go. He also talked about the challenge to provide people ample warning during severe weather. Dolusic compared doing the weather during a tornado outbreak to trying to simultaneously cover both Ole Miss and Mississippi State games if they were playing at the same time. "You can't give full coverage to each one, but you have to try to," Dolusic said.

Bassett recalled Hurricane Katrina as her most memorable moment. On the professional side, she recalled needing to convince her station to lead the newscast with her Katrina forecast on the Friday before it made landfall because it was the first time it looked as if the hurricane would impact the Mississippi coast. Meanwhile, on a more personal level, Bassett was pregnant during the hurricane. She recalled how difficult it was to see all the people who had to take shelter in the Coliseum—especially all the women who were pregnant in a new place without even knowing where they would deliver their babies. Fortunately, Bassett was able to arrange "Jackson's Biggest Baby Shower" for the evacuees who were expecting and was able to serve 80 women by giving them baby supplies they needed and introducing them to local doctors and hospitals.

WHAT YOU MAY NOT KNOW ABOUT YOUR LOCAL WEATHERCASTER

WLBT meteorologist Barbie Bassett has published two books, including *Forecasts and Faith: Five Keys to Weathering the Storms of Life*. When asked about the relationship between her faith and her career in television weathercasting, Bassett said, "Forecasting and faith are very similar in that you're taking things that can't be seen and you're making them into things that can be seen."

WTVA meteorologist John Dolusic's father was taken as a prisoner of war by the Nazis while he was living in Croatia during World War II. His mother was also from Croatia. Initially, his parents hoped he would be a doctor or a lawyer, but he was more interested in weather. Dolusic said that it is nice to live in a country "where folks can still come to the United States and make a dream, if not for themselves, to make a dream for their generations to come."

Former WCBI meteorologist Jason Dunning's favorite part of his job is getting to share his passion for weather with others, something he has been doing since he was a little boy. When he was a child, Dunning would tape the weather forecasts on the VCR, and then take his mother's video camera and record himself pointing at the maps and telling the weather story like the people he watched on television.

Finally, these weathercasters were asked what advice they have for an aspiring television meteorologist. Dolusic said if a person loves weather, he or she should go into weather, but if someone wants to be on television, that person should be an actor instead. You are a forecaster first. He also urged those interested to "be yourself" and "do your best." Similarly, Dunning said that students should have a "very strong passion and love for weather." He also urged future television meteorologists to pursue this career for the love of science and not just to be on television. Dunning also added that future forecasters need to be prepared to not make a lot of money. Bassett and Dolusic agree that the business is also competitive. Bassett's advice is "be ready for change." Television changes by the year, and jobs are not as numerous as they once were, so it will help the aspiring television weathercaster to know how to do as many things as possible at the station, such as editing, shooting video, producing, communicating via social media, and gathering information.

WEATHER MYTHS AND FOLKLORE

Before we had television and internet to give us weather forecasts, people often observed interesting natural sights that could help predict the weather. Some are common to many places, while others are more specific to the Southeast. Weather sayings that have the most credence are those that involve sky conditions that can be easily observed. For example, two sayings involving the night sky both refer to the moisture content of the atmosphere, which is a good predictor of weather.

The saying "Halo around the moon, rain or snow will be coming soon" is often true, as is the saying "Cold is the night when the stars shine bright." The halo around the moon is the result of ice crystals in the atmosphere bending the light. You might think that ice crystals should mean that snow would be coming, but ice crystals are always present in the highest type of cloud—cirrus clouds. Cirrus clouds often indicate the approach of a warm front, which is often accompanied by rain. The snow part of the saying is more applicable in other parts of the country, where warm air brings the moisture needed for snowfall. Similarly, the stars are brightest when there are no clouds present—cirrus or otherwise. Cloudless nights most often occur when a place is under high pressure. The coldest nights are also typically after a cold front passes by and the area is under high pressure with no winds.

Another familiar saying, "Red sky at night, sailor's delight; red sky in morning, sailors take warning," is also true. This phrase has two parts. The red sky at night is caused by dust particles in the atmosphere bending the light from the setting sun. High pressure allows the dust particles to stay suspended in the air. Because the sun sets in the west, this means that the high pressure is also in the west. High pressure brings fair weather, hence the sailor's delight. On the other hand, if the sky is red in the east while the sun is rising, it means that the fair weather and high pressure has passed. Clouds and low pressure may be moving in. This is true for areas of the United States where the weather typically moves from west to east. It would not be as applicable for areas at polar or tropical latitudes. Interestingly, in some form or another this is one of the oldest weather expressions. Sky color and weather is even referenced in the Bible. Matthew 16:2–3 states "When evening comes, you say, 'It will be fair weather, for the sky is red,' and in the morning, 'Today it will be stormy for the sky is red and overcast.'"

Another class of weather predictors involves the behaviors or presence of certain living things. There are a number of weather myths related to woolly caterpillars, also called woolly bears or woolly worms, which may have first been told by Native Americans. The predominant myth is that the thickness of the rust stripe on the woolly bear caterpillar can predict how harsh the winter will be—the thinner the band, the more severe the winter. One scientist suggested that a relationship might exist, but it would not be very useful in predicting weather. The width of the stripe might be related to conditions in the *previous* winter. Other weather myths related to the woolly bear include the thickness of their furry coat (the thicker the coat, the harsher the winter), the direction they are moving (moving north indicates a milder winter than moving south), or their abundance (more woolly caterpillars foretells a hard winter).

Another winter weather prognosticator that seems to be most common in the southern states is the persimmon seed. The myth says that when you cut a persimmon and split its seeds down the middle, they will reveal the shape of a knife, a spoon, or a fork. If you see the knife, winter will be cold (the cold weather will cut you like a knife). If the seed displays a fork, the winter will be mild. The spoon indicates a snowy winter may be ahead. The problem with this myth is that the spoon seems to be the most common shape regardless of where, or in what year, the seeds are split.

There is other animal lore that can help you predict the weather, although unlike the woolly bear myth, these associations deal with weather that is either occurring or about to occur. For instance, birds often fly lower before a storm. This is due to a drop in air pressure before the rain. Dogs who are afraid of thunderstorms will often know a storm is coming before his owner. The frequency of a cricket chirp is actually related to the temperature. There may even be a bit of predictive power in the direction a cow is standing. According to the North Carolina State Climatologist, if the cow stands with its tail toward the west, fair weather is coming. If the cow stands with its tail facing east, foul weather is ahead. Cows stand with their tails toward the wind. Winds from the east bring rain, while westerly winds usher in fair weather.

The Native Americans who lived in Mississippi also had weather-related stories. The Choctaw had a story describing the origin of thunder and lightning. According to Choctaw mythology, they were caused by two large birds, Heloha (thunder) and Melatha (lightning). Heloha's eggs would roll around the clouds where she laid them and make the rumbles of thunder. Melatha caused the lightning when he made sparks as he raced across the sky to try to catch the eggs.

SCIENTIFIC WEATHER FORECASTING

Like weather-related folklore, the scientific measurement and prediction of weather also has a long history. Individuals have only been acquiring weather forecasts from television since the late 1940s, but people have been observing weather phenomena since before the days of Aristotle, who wrote *Meteorologica*, the book containing his theories about the Earth, in 350 B.C.

In his book *Weather Studies*, Joseph Moran (2006) shared highlights in weather history. The oldest entry is from 525 B.C., when a Greek philosopher proposed that air thickened to make clouds, wind, and rain. Other notable entries include the first rain gauge used in India around 400 B.C.; Galileo's 1592 invention of the thermoscope, which precede the thermometer; Torricelli's 1643 invention of the mercurial barometer; Benjamin Franklin's electricity experiments, which began in 1747; the first weather map in 1819; the first use of radar

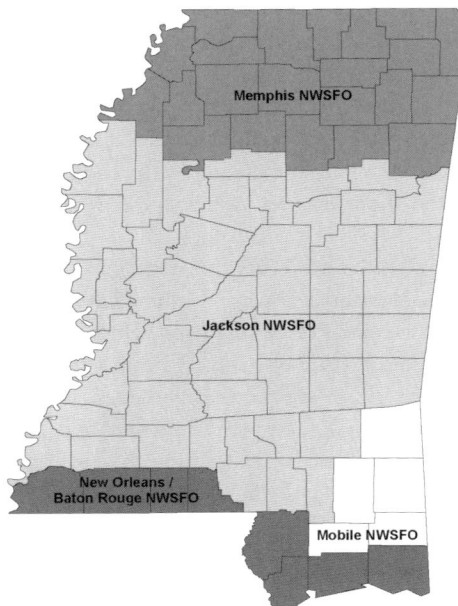

Fig. 10.3. National Weather Service county warning areas in Mississippi. The only office based in Mississippi is the Jackson National Weather Service forecast office.

to track a weather system in 1947; the first tornado warning in 1948; and the launch of the first weather satellite, TIROS I, in 1960. There are many more important events in weather history, but they are too many to list here. One of the most important advances for the United States was the development of what is now known as the National Weather Service (NWS).

THE NATIONAL WEATHER SERVICE

The state of Mississippi is divided among four NWS forecast offices, but only one of these is headquartered in Mississippi (Figure 10.3). The Jackson NWS forecast office forecasts for 47 of Mississippi's counties, as well as additional counties in Louisiana and Arkansas. The remaining counties are covered by the Memphis office, which forecasts for 22 counties in northern Mississippi, and the New Orleans and Mobile offices, which share the remaining counties in southern Mississippi.

The NWS had its origins in the military. President Ulysses S. Grant signed a joint resolution creating what would become today's NWS in 1870. The resolution required the Secretary of War "to provide for taking meteorological observations at the military stations in the interior of the continent and at other points in the States and Territories . . . and for giving notice on the northern

Fig. 10.4. The cotton region shelter used by some cooperative observers. The shelter holds thermometers and other weather instruments. Photo credit: National Oceanic and Atmospheric Administration, Department of Commerce

[Great] Lakes and on the seacoast by magnetic telegraph and marine signals, of the approach and force of storms" (NWS Public Affairs Office 2010). This early weather service was placed under the Secretary of War because it was thought that this would ensure accuracy and regularity. Beginning on November 1, 1870, at 7:35 a.m., reports were taken at the same time at 24 Army Signal Service Stations throughout the country. Initially, forecasts were made only for eight districts that covered the whole country. The length of advanced prediction increased gradually, but the number of observation stations increased by an order of magnitude in the first decade. The use of "forecast" was actually not sanctioned until 1899. Prior to that, weather forecasts were called "probabilities" or "indications."

The Signal Service began coordinating weather observations for the benefit of the cotton industry in 1881. The Cotton Region Observers recorded weather information during the growing season across the southeastern United States. Their reports were telegraphed to New Orleans and then shared at cotton exchange newspapers and bulletins (Grice 2006c).

According to the annual report of the Chief Signal Officer (Fulton 1892), there was an attempt made in 1885 to organize a corps of voluntary observers in Mississippi. Efforts to get the state legislature to equip volunteers with instruments were unsuccessful, despite two special messages from the governor. Some thermometers were supplied by the Signal Service in 1887, and the University of Mississippi provided an office and the costs associated with publishing monthly bulletins. The reports of the observers were incorporated with the Signal Service station reports from Vicksburg and Meridian and other stations

Fig. 10.5. Sign outside of Jackson forecast office of the National Weather Service. Photo credit: K. Sherman-Morris

outside of Mississippi. Weekly weather crop bulletins were credited with generating an interest in weather, especially in frost warnings. Observations were made with the help of a cotton region shelter (also called a Stevenson screen after its inventor Thomas Stevenson; Figure 10.4), which was a white box, typically of pine construction with louvered sides, that contained thermometers and other weather instruments. It was mounted 4–6 feet off the ground on a four-legged stand.

In 1890, the weather personnel from the Signal Service were given an honorable discharge from the War Department and the Weather Bureau was transferred to the Department of Agriculture by President Benjamin Harrison. For the first time, forecasters would be civilians. In 1970, 100 years after its founding, the Weather Bureau officially became the NWS, which had recently been incorporated by the Department of Commerce, its current home.

Before 1931, observations were made by individuals working with the Smithsonian Institution, Signal Service/Weather Bureau Cotton Region, Army Medical Department, and cooperative observing programs. Observations were not made at Weather Bureau stations until 1931 (Jackson NWS website). That year, observations were recorded for Jackson at Hawkins Field, the site of the newly opened Weather Bureau Airport Station. This office was closed in 1935, a few years after opening, due to budget restriction. The New Orleans office made the daily forecasts for Mississippi, beginning when the Jackson office closed from 1935 to 1939 and continuing through the 1960s. Memphis and New Orleans shared forecasting responsibilities from the 1960s until 1972. From 1940 until 1972, the Jackson office issued mostly weather warnings and adjustments, and they continued to make observations.

In 1963, the Jackson Weather Bureau Office relocated to the location of the new airport east of Flowood, built to serve the area's growing air transportation needs. The Jackson Weather Bureau had been launching weather balloons since 1953 and using radar since 1959. Even equipped with these tools, it was not until

Fig. 10.6. Balloons (left) are inflated and sent up into the atmosphere from buildings at National Weather Service forecast offices (below). Photo credit: K. Sherman-Morris

A DAY IN THE LIFE OF A NATIONAL WEATHER SERVICE FORECASTER

Joanne Culin is a forecaster at the NWS forecast office in Jackson. Like many meteorologists, she became interested in weather after experiencing extreme weather—in her case, the 1995 hurricane season. Culin described a typical day working at the NWS. The meteorologists work on a staggered shift schedule, and there are two forecasters working on each shift. Day shift is typically the busiest. One forecaster is responsible for providing weather updates and forecasts for the airport, while the other works on forecasts for the public. Other duties include watching the radar and tracking any developing hazards. During slower shifts, such as the evening shift, meteorologists can work on projects to more closely study specific aspects of the weather.

Culin also described how she makes a forecast. This involves looking at the current conditions, including what the satellite shows, and what major weather features exist at different levels in the atmosphere and then comparing the way different computer models are handling those features. She determines which model is depicting the current conditions the best and if any of the models show the ingredients for severe weather or winter weather. Graphical forecast editing software is used to draw a representation of the forecast. Culin said that dealing with complex severe weather events are the most challenging aspect of her job. During big weather events, more people are on hand to help with various duties, including monitoring radar, communicating with emergency managers, and issuing warnings. Sometimes ham radio operators even come to the NWS office to monitor severe weather reports from the field and transmit information.

Culin said that, while the job can be difficult at times, it can also be very rewarding when information in a warning helps to save lives, as it did in the April 2010 Yazoo City tornado. To young people interested in a career in meteorology, she suggests, "If it's something you're passionate about, don't give up because of all the requirements you have to take in school."

Fig. 10.7. Joanne Culin demonstrating the graphics forecast editor. Photo credit: K. Sherman-Morris

Fig. 10.8. The inside of the Jackson National Weather Service forecast office. Lots of computers are required to track the changing weather. Photo credit: John Morris

April 1972 that the Jackson office was charged with issuing forecasts for most of Mississippi. By this time, the name of the Weather Bureau had changed to the National Weather Service. The current responsibilities for issuing forecasts and warnings lie with the NWS forecast offices shown in Figure 10.2. The Jackson NWS office got its own building at the airport in 1978 and has continued to add more sophisticated forecasting tools and equipment as new technology has become available.

In 2010, the Jackson NWS employed 12 full-time forecasters as well as more than a dozen other professionals. Three individuals form the management team for the office. This includes the warning coordination meteorologist, who is the liaison between the NWS and user of the NWS products as well as the coordinator of all public education efforts; the science and operations officer, who is in charge of researching and implementing new scientific technologies, including training staff members; and the meteorologist in charge, who manages and directs the activities at the office. The office has many tools to help create the forecast.

WEATHER OBSERVATION NETWORKS

From the beginning of officially recorded weather observations, the observations were made by a number of individuals, some affiliated with the Signal Office or Weather Bureau and some affiliated with other organizations. Other people simply kept records for their own interest. The first observations made in Mississippi were largely voluntary. The Station Histories Project, initiated by the National Oceanic and Atmospheric Administration's (NOAA) National Climatic Data Center (NCDC), commissioned Gary K. Grice to collect information about early weather observations made in Mississippi. Grice has written histories of weather observations in four cities: Columbus, Jackson, Natchez, and Vicksburg. These histories are available through the Midwestern Regional Climate Center (Grice 2006a, 2006b, 2006c, 2006d). The description that follows is largely summarized from these histories.

The City of Columbus, Mississippi, was founded in 1821 after being settled in 1817. Unlike Jackson or some of other cities, Columbus did not have a Signal Service or Weather Bureau office, but other organizations that kept early weather observations were active in Columbus. One weather observation program was conducted by the Smithsonian Institution, which began to provide weather instruments to observers in the 1840s as a way to help detect storms across the country (Fiebrich 2009). Soon, hundreds of observers were at work, including several in Columbus. One of the most complete weather records kept for the Smithsonian in Columbus was made by James S. Lull from 1856 to 1871 (Grice 2006a). It was also the first record from Columbus in the NCDC database. Lull took weather observations from his downtown home, which was and still is known as Camellia Place. Other observers kept a record for a couple years at a time.

Another common source of early weather observations was the U.S. Army Medical Department. In 1814 the surgeon general of the Army required all U.S. Army hospital surgeons to keep a diary of the weather, and in 1819 this was made a regular practice of army posts (Fiebrich 2009). An observation of only one year (1868–1869) was kept in Columbus by army field surgeons, and the location of these observations is unknown (Grice 2006a).

From 1872 to 1883 no records were kept in Columbus (Grice 2006a). A series of three observers recorded the weather at the Mobile and Ohio Railroad Depot in Columbus under the auspices of the Cotton Region Observers from 1883 to 1896. After that, the station became part of the Weather Bureau's Voluntary Observing Program (Grice 2006a). A number of other observers also became volunteers under this program.

The Smithsonian Institution program was also responsible for the first official weather observations recorded in Jackson. These were taken by young women enrolled at the Oakland (Ladies) Institute as part of their training in the arts and sciences (Truesdell et al. 2007). The observations were kept in their Meteorological Journal from 1849 to 1852 and possibly through 1854 (Grice 2006b). Other early observations for the Smithsonian program were made at the Jackson Female Institute (1853–1854) and Mississippi College in Clinton (1870–1871) (Grice 2006b). The length of observations taken by the U.S. Army surgeons was longer in Jackson than it had been in Columbus. These observations were made from 1873 to 1876, which was still not long enough to prevent a gap in the records for the Jackson area; the next set of observations were not made until 1883, when the Cotton Region observations began (Grice 2006b).

The record of observations taken by the Cotton Region observers was pretty well covered from 1884 to 1893, when the program was taken over by the Voluntary Observer Program. One of the longest records of observations was taken at the Alabama and Vicksburg Depot on Court and State Streets in Jackson from 1883 until 1899 (Grice 2006b). Observations were also taken at the Western Union Office from 1899 to 1905. Once the Voluntary Cooperative Observer Program began, observations were taken on a more continuous basis at various locations around Jackson. Several voluntary observers kept weather records from the 1890s to the 1930s. The NWS continues to use cooperative observers to provide temperature and precipitation data.

Weather observations were first recorded (in French) in Natchez in 1795, and regular and systematic observations were kept from 1799 to 1818 by William Dunbar and his son at their plantation, The Forest, located south of Natchez (Grice 2006c). Along with temperature, pressure, rainfall, wind, and the state of the weather, the Dunbars often wrote notes on each monthly observing form noting the progress of a particular crop or the birth of calves (Grice 2006c). Other plantations may have kept weather records in diaries during the early 1800s. Several Smithsonian observers recorded for Natchez in the mid-19th century, and these observers helped to keep the record fairly well covered from 1849 to 1870. Some of these observers made very careful notes about the quality and use of their weather instruments. As in Columbus, the U.S. Army surgeons only kept weather records for about a year. Cotton Region and later cooperative observers continued to maintain an almost continuous record of the weather in Natchez from 1884 to 1911. Two families accounted for 42 years (1911–1953) of observations from downtown Natchez: the Garrity and the Butchart families, who were related (Grice 2006c).

Vicksburg was somewhat unique in Mississippi in that it had a Signal Service office. Observations were taken there from 1871 until 1891, when the Weather Bureau took over. Some evidence suggested that official weather observations began in Vicksburg as early as 1840; however, early records may have been destroyed in the 1960s when the office closed (Grice 2006d). The first official weather observation made in Vicksburg was recorded in March 1849 by a Smithsonian Institution observer. These observations were made until 1852

Fig. 10.9. Miss Annette Koch was an observer at Pearlington, Mississippi, in the late 19th and early 20th centuries. The picture was taken in 1935. Photo credit: National Oceanic and Atmospheric Administration, Department of Commerce

(Grice 2006d). The record of observations made by U.S. Army surgeons ran from 1866 to 1870. Weather Bureau observations were made from the Post Office building at Walnut and Crawford Streets in downtown Vicksburg from 1891 to 1955, with their instrument shelter and other instruments located on the roof until 1818 (Grice 2006d).

Other known weather observations were made from the mid-1800s to the early 1900s in Fayette, Stonington, Roxie, Suffolk, Knoxville, Woodville, Wilkinson, Briers, and Washington (Grice 2006c). Many of these stations kept records for a long time. For instance, almost continuous observations were made in Woodville from 1893 to 1955, where stations had a cotton region instrument shelter, thermometers, and rain gauge (Grice 2006c). Similarly impressive was the 33-year record in Fayette (Grice 2006c). The Signal Service catalogued 10 observers throughout Mississippi in the 1850s, but that number grew to 43 in the 1880s (Fiebrich 2009).

CoCoRaHS

The Community Collaborative Rain, Hail, and Snow Network (CoCoRaHS) is a volunteer network of people interested in the weather who report daily rainfall observations. In 1997, a devastating flood hit Fort Collins, Colorado, causing $200 million in damages. The rainfall that caused the flash flood was very

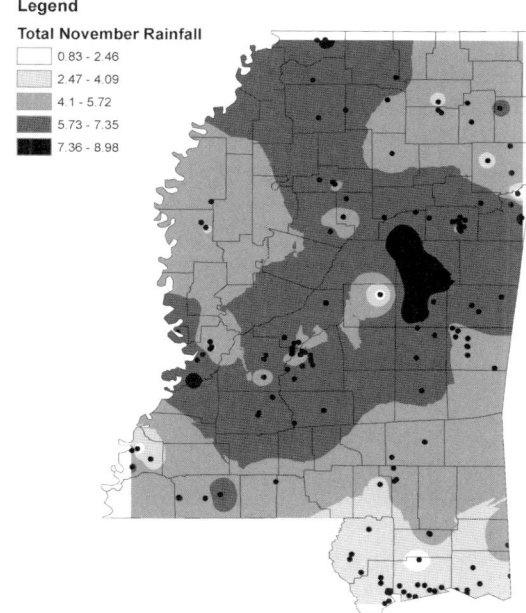

Legend

Total November Rainfall

- 0.83 - 2.46
- 2.47 - 4.09
- 4.1 - 5.72
- 5.73 - 7.35
- 7.36 - 8.98

Fig. 10.10. Rainfall observations reported during November 2010 by Mississippi CoCoRaHS observers

localized, ranging from less than 2 inches to more than 14 inches of rainfall in a little over a day, with most of the heaviest rainfall coming in less than 6 hours. Initially, the actual rainfall amounts were unknown because of the lack of official rain gauges. The magnitude of the rainfall even surprised meteorologists. The assistant state climatologist at the time, Nolan Doesken, began to collect rainfall observations from citizens around Fort Collins and eventually determined the rainfall patterns. This incident illustrated the importance of a dense network of rainfall observations, and in 1998 CoCoRaHS was born.

Mississippi joined CoCoRaHS in 2008. At the end of 2010, the state had about 325 observers. Mississippi CoCoRaHS is always looking for new observers. To participate, a volunteer has to purchase an official rain gauge, install the gauge somewhere he or she will read it regularly, and then report the rainfall amount on the CoCoRaHS website (www.cocorahs.org) each day, preferably between 6:00 and 9:00 a.m. Observers also sometimes make comments along with their reports. These may be about how heavy the rain was, when precipitation occurred, how bad the drought was getting, or comments unrelated to precipitation such as remarks about the temperature or other conditions.

In December 2009, an observer from Adams County reported, "First time to ever report snow! Wow. How pretty it all is." Following heavy rain that produced more 5 inches in only 4 hours in July 2011, an observer from Harrison County reported, "A lot of rain since approximately 4:00 AM. No flooding found due to such dry conditions for most of the year but huge puddles where streets do not drain very well!" Two comments from an observer in Warren County from August 2010 provide some insight into how people feel about the weather and how quickly the weather can change. On August 16 the observer wrote, "Not a lot but at this point we'll take what we can get. We were beginning to get quite dry here." By August 19, he had the following to say, "A Tropical Low is to blame for our beneficial rain lately. Now it can move on out as far as I'm concerned."

CoCoRaHS has definitely contributed to our record of precipitation events in the state. This information would be even better with greater coverage across the state. Figure 10.10 shows the variability across the state of total rainfall during a month. Each dot represents a CoCoRaHS reporting station. You can see from the map that observers are not evenly distributed throughout the state.

METEOROLOGY EDUCATION IN MISSISSIPPI

Meteorologists are employed by the NWS and by television stations. Although many meteorologists are employed by government agencies, several private companies also exist to provide weather forecast information. Meteorologists are also hired to provide forecasts for a variety of private businesses, such as supermarket chains, airlines, and power companies.

A bachelor's degree is required for most meteorology careers, whereas some research positions at universities or government agencies require a Ph.D. Students wishing to pursue careers in meteorology can obtain a degree in meteorology or in a related field. Some employers, such as the NWS, have specific course requirements in science and math. There are two schools in Mississippi that prepare students for careers in meteorology, Mississippi State University in Starkville and Jackson State University in Jackson.

Mississippi State University

Mississippi State University began as Mississippi Agricultural and Mechanical College in 1878. In the very early days of the college, meteorology was actually required of all students and it was offered by the Chemistry Department. In 1905 the course was moved to the Geology Department, which became the Department of Geology and Geography in 1936. For a time, meteorology was not taught. When climatologist Charles Wax joined the department in 1977, meteorology was once again a part of the curriculum. Meteorology courses were added during the 1980s and 1990s, and the Broadcast Meteorology Program to train students to become weathercasters was developed. Eventually, this program expanded through distance learning to reach television weathercasters throughout the country as well as those outside the United States.

Today the Department of Geosciences is home to more than 30 faculty members, including 15 faculty with a specialty in meteorology or climatology. In Fall 2009 the Department of Geosciences was home to more than 125 undergraduate students and 55 graduate students, many of whom were also concentrating in a weather- or climate-related field. Twenty-eight meteorology courses are regularly taught, including weather forecasting and broadcast meteorology. The department offers a B.S. and M.S. in Geosciences as well as a Ph.D. in Earth and Atmospheric Sciences.

Graduates from the meteorology program in the Department of Geosciences at Mississippi State have gone on to work at the NWS and the Weather Channel and are employed in every market in the state of Mississippi, as well as many other markets throughout the country. Students have also been quite successful at forecasting. In 2008–2009, 2009–2010, and 2010–2011 the weather forecasting team at Mississippi State took first place in WxChallenge, the North American Collegiate Weather Forecasting Competition. During the 2009–2010 school year, for example, the Mississippi State University team was ranked first, ahead of 45 other schools.

Jackson State University

A B.S. in Meteorology is offered by the Department of Physics, Atmospheric Sciences, and General Sciences at Jackson State University, which is the nation's only historically black college or university to offer a bachelor's degree in meteorology. The program in meteorology was initiated by the school's president, John A. Peoples, in 1975 and authorized the following year. The first degree in meteorology was granted in 1980.

Many of the students who graduate from Jackson State go on to work for NOAA. A smaller number work in television weathercasting, including both local markets and the Weather Channel, or go on to teach or work in the private sector. Many students also choose to pursue graduate degrees. The program at Jackson State University provides students many opportunities to get involved with research. Students have completed research-oriented internships with the National Center for Atmospheric Research, the National Hurricane Center, and other NOAA offices. The Department of Physics, Atmospheric Sciences, and General Sciences offers 13 meteorology classes and is home to more than 20 faculty members, including three faculty members with a specialty in meteorology.

Further Reading

For information about the people discussed in this chapter, see the following websites:

Barbie Bassett: http://www.wlbt.com/Global/story.asp?S=213233

John Dolusic: http://www.wtva.com/news/staff/jdolusic.shtml

David Hartmann: http://www.wapt.com/newsteam/739181/detail.html#

Mike Reader and Tommy Richards: http://www.wlox.com/Global/category.asp?C=2606

Dick Rice: http://www.wtva.com/news/staff/drice.shtml

NWS Jackson: http://www.srh.noaa.gov/jan/

NWS Memphis: http://www.srh.noaa.gov/meg/

For more information about the history of television weathercasting, see:

Henson, R. 1990. *Television Weathercasting: A History*. Jefferson, NC: McFarland and Co.

For more information about the meteorology programs at Mississippi State and Jackson State, see:

Jackson State University Department of Physics, Atmospheric Sciences and General Sciences: http://msp.jsums.edu

Mississippi State University Department of Geosciences: http:// geosciences.msstate.edu

Afterword

THE FLOOD OF 2011 ON THE MISSISSIPPI RIVER

The year 2011 was a year of extremes in weather in many parts of the U.S., and the impact on Mississippi was severe. The historic Mississippi River flood of 2011 is a good example of how weather events all across the continent can affect the state. Record snow amounts fell across the northern parts of the Mississippi River Basin during the winter months then began melting in March. Continuous heavy rainfall joined the melting snow at the end of April over the same northern parts of the Mississippi River Basin, north Arkansas, south Missouri, and the Ohio River Valley. Over the two-month period from April 1st to June 1st, a total of 30 to 35 inches of rain (300%–400% of normal) fell over parts of southern Missouri. All that water had nowhere to go except down the Mississippi River, past Mississippi, to the Gulf of Mexico.

The Mississippi River had begun following its normal annual pattern—it rose in the early spring, crested in the middle of April, then began to recede. Not surprisingly, given the hydrologic situation in the basin above Mississippi, the River began to rise again at the end of April. Heavy rain continued in the river's basin north of Mississippi in May. That heavy rain combined with the earlier rainfall in the same locations and the melting snow from further north produced record or near-record flooding along the river as it flowed past Mississippi. This surge of water pushed the flood's crest past the 1937, 1973, and 2008 flood levels at Arkansas City and Greenville, and even to record flood crests at Vicksburg (57.1 feet—previous record 56.2 feet in 1927) and Natchez (61.9 feet—previous record 58.04 feet in 1937). The backwater areas up the Yazoo River crested above all previous records except 1927 levels, and were 6 to 7 feet higher than 2008 river levels. The river finally crested in mid May, with most points not falling below flood stage until the first and second weeks of June. All told, the Mississippi River was above flood stage for nearly two and a half to three months as shown in the flood stage report below.

The flood was remarkable in many historic ways, and strange in others. For example, on a field trip to Stoneville, Mississippi, on May 19 when the flood was at its height just about eight miles away, dust covered vehicles on the gravel roads and in the fields. Flooding and flood damage was limited to the area inside the levees and to backwater flooding where tributaries could not flow into the Mississippi River. In this sense, the levees saved most of Mississippi from

Location	Flood Stage (FS) in feet	Dates Above FS	Crest—date
MISSISSIPPI RIVER			
Arkansas City, MS	37	4/28—6/11	54.14—5/16
Greenville, MS	48	4/28—6/13	64,22—5/17
Vicksburg, MS	43	4/30—6/17	57.10—5/19
Natchez, MS	48	4/29—6/22	61.95—5/19
YAZOO RIVER			
Yazoo City, MS	18	5/03—6/12	38.7—5/21
(Data from NWS Form E-5, Jackson, MS Hydrologic Service Area, July 27, 2011)			

the ravages of the historic flood, unlike the 1927 flood when most of the Delta and other areas along the River were under water.

Even though the levees performed their assigned task, the flood still had a significant impact on people, property, and wildlife not contained within the levee system. According to the National Weather Service in Jackson, Mississippi, more than 1500 buildings were damaged or destroyed. The Mississippi River flood was responsible for major damage to 40 residences in Vicksburg and Warren County where there was no levee protection, and backwater flooding in Yazoo and Humphreys Counties caused major damage to at least 50 residences. An additional 60 homes in Washington County were destroyed.

Economic losses from the flood of 2011 all along the Mississippi River range from 2–4 billion dollars. In Mississippi, where nearly 400,000 acres were under water, The National Oceanic and Atmospheric Administration estimates damage to agriculture of $800 million. Fourteen counties were declared major disaster areas by President Obama. According to the Federal Emergency Management Agency, close to $20 million in state or federal assistance was approved for flood survivors in Mississippi and 100,000 cubic yards of debris was removed. Allen Godfrey, the deputy director of the Mississippi Gaming Commission stated the closures of the 17 casinos due to the flood cost the state about $7.2 million in lost taxes (Sayre, 2011).

Further Reading

Jackson National Weather Service Forecast Office (2011). Historic Mississippi River Flood of 2011, Available online at http://www.srh.noaa.gov/jan/?n=2011_05_ms_river_flood

National Oceanic and Atmospheric Administration (2011). Billion Dollar U.S. Weather Disasters, Available online at http://www.ncdc.noaa.gov/oa/reports/billionz.html

Sayre, A. (July 21, 2011). Miss. Casino Revenue Makes Post-Flood Comeback. The Memphis Daily News 126, Available online at http://www.memphisdailynews.com/news/2011/jul/21/miss-casino-revenue-makes-post-flood-comeback/

BIBLIOGRAPHY

Chapter 2

Blake, E. S., E. N. Rappaport, and C. W. Landsea. 2007. *The Deadliest, Costliest, and Most Intense United States Tropical Cyclones from 1851–2006*. NOAA Technical Memorandum NWS TPC-5. http://www.nhc.noaa.gov/pdf/NWS-TPC-5.pdf, accessed May 12, 2011.

Brown, M. E., and C. Wax. 2007. Temperature as an Indicator of Landform Influence on Atmospheric Processes. *Physical Geography* 28:148–157.

Dyer, J. L. 2010. Four-Dimensional Visualization and Analysis of Convective Rainfall Generation along an Abrupt Land Use/Land Cover Boundary in Northwest Mississippi. Paper presented at the 90th Annual Meeting of the American Meteorological Society. http://ams.confex.com/ams/90annual/techprogram/paper_160024.htm, accessed December 14, 2010.

Lutgens, F. K., and E. J. Tarbuck. 2004. *The Atmosphere: An Introduction to Meteorology.* 9th ed. Upper Saddle River, NJ: Prentice Hall.

Moran, J. M. 2006. *Weather Studies: Introduction to Atmospherics Science.* 3rd ed. Boston, MA: American Meteorological Society.

Robinson, P. J., and A. Henderson-Sellers. 1999. *Contemporary Climatology.* 2nd ed. Harlow, England: Pearson Education Limited.

U.S. Naval Observatory. The Dark Days of Winter. http://aa.usno.navy.mil/faq/docs/dark_days, accessed December 8, 2008.

Chapter 3

Holder, C., R. Boyles, P. Robinson, S. Raman, and G. Fishel. 2006. Normal Temperature Range that Reflects Daily Temperature Variability. *Bulletin of the American Meteorological Society* 87: 769–774.

Lupo, A. R., E. P. Kelsey, E. A. McCoy, C. Halcomb, E. Aldrich, S. N. Allen, et al. 2003. The Presentation of Temperature Information in Television Broadcasts: What Is Normal? *National Weather Digest* 27: 53–58.

Chapter 4

Barry, J. M. 1997. *Rising Tide: The Great Mississippi Flood of 1927 and How It Changed America.* New York: Touchstone.

Burt, C. 2004. *Extreme Weather: A Guide and Record Book.* New York: W. W. Norton and Co.

Edelen, G. W., Jr., K. V. Wilson, and J. R. Harkins. 1986. *Floods of April 1979: Mississippi, Alabama, and Georgia.* U.S. Geological Survey/National Oceanic and Atmospheric Administration, U.S. Geological Survey Professional Paper 1319. Washington, D.C.: U.S. Government Printing Office. http://choctaw.er.usgs.gov/new_web/reports/other_reports/flood/flood79.html, accessed December 15, 2009.

Frankenfield, H. C., M. W. Hayes, H. S. Cole, W. E. Barron, R. T. Lindley, F. W. Brist, et al. 1927. The Floods of 1927 in the Mississippi Basin. *Monthly Weather Review*, Supplement 29 (October 18).

Hederman, T. M. 1979. *The Great Flood.* Jackson, MS: Clarion Ledger/Jackson Daily News.

Henry, A. J. 1927. Frankenfield on the 1927 Floods in the Mississippi Valley. *Monthly Weather Review* 55: 437–452.

National Weather Service. 2008. Southeast U.S. High Fire Danger Weather Patterns. http://www.srh.noaa.gov/jan/?n=seus_fire_weather_favorable_pattern, accessed December 5, 2008.

National Weather Service. n.d. April Pearl River Flood of 1979: Easter Flood of 1979. http://www.srh.noaa.gov/jan/?n=1979_04_17_easter_flood_chronology, accessed December 15, 2009.

Public Broadcasting Corporation. n.d. Fatal Flood: Timeline. http://www.pbs.org/wgbh/amex/flood/timeline/index.html, accessed December 15, 2009.

Time Magazine, no author listed. Catastrophe: Deluge. May 2, 1927. http://www.time.com/time/magazine/article/0,9171,751651,00.html, accessed December 15, 2009.

Chapter 5

Brooks, H., and C. Doswell. 2001. Deaths in the 3 May 1999 Oklahoma City Tornado from a Historical Perspective. *Weather and Forecasting* 17: 354–361.

Brown, M. E., and C. L. Wax. In press. *Thunderstorms, Lightning Strikes, and Tornadoes in Mississippi.* MAFES Bulletin, Mississippi Agricultural and Forestry Experiment Station, Mississippi State University.

Grazulis, T. 1993. *Significant Tornadoes 1680–1991.* St. Johnsbury, VT: The Tornado Project of Environmental Films.

Gunn, A. M. 2008. *Encyclopedia of Disasters: Environmental Catastrophes and Human Tragedies.* Westport, CT: Greenwood Press.

Johnson, D., G. Holly, and C. Kieffer. 2011. Smithville Deaths at 13, Recovery Begins. *Northeast Mississippi Daily Journal,* A1.

Kelly, D., J. Schaefer, and C. Doswell. 1985. Climatology of Nontornadic Severe Thunderstorm Events in the United States. *Monthly Weather Review* 113: 1997–2014.

LeCoz, E. 2011. Survivors Recall the Twister. *Northeast Mississippi Daily Journal,* A1.

National Oceanic and Atmospheric Administration (NOAA). 2010. Storm Data Events. http://www4.ncdc.noaa.gov/cgi-win/wwcgi.dll?wwEvent~Storms, accessed June 15, 2010.

National Weather Service, Memphis. 2011. Public Information Statement. http://www.srh.noaa.gov/meg/?n=lateapr2011outbreak, accessed May 16, 2011.

Texas Tech University. 2006. *A Recommendation for an Enhanced Fujita Scale (EF Scale).* Lubbock: Wind and Science Engineering Center, Texas Tech University.

Chapter 6

Ashley, S. T., and W. S. Ashley. 2008. Flood Fatalities in the United States. *Journal of Applied Meteorology and Climatology* 47: 805–818.

Bellande, R. L. 2005. Neighborhoods. http://oceanspringsarchives.net/node/142, accessed November 4, 2010.

Dunn, G. E., and B. I. Miller. 1960. *Atlantic Hurricanes*. Baton Rouge: Louisiana State University Press.

Elsner, J. B., and A. B. Kara. 1999. *Hurricanes of the North Atlantic: Climate and Society*. New York: Oxford University Press.

Hancock County Historical Society. n.d. Storms. http://www.hancockcountyhistorical society.com/history/disasters.htm#storms, accessed November 4, 2010.

Keim, B. D., and R. A. Muller. 2009. *Hurricanes of the Gulf of Mexico*. Baton Rouge: Louisiana State University Press.

National Oceanic and Atmospheric Administration (NOAA). 2007. From Kites to Satellites: A History of Weather and Air Research at NOAA. http://celebrating 200years.noaa.gov/foundations/atmospheric/welcome.html#hurricane, accessed May 13, 2011.

National Oceanic and Atmospheric Administration, National Weather Service, and National Hurricane Center. 1993. Memorable Gulf Coast Hurricanes of the 20th Century. http://www.aoml.noaa.gov/general/lib/mgch.html, accessed November 4, 2010.

Rappaport, E. 2000. Loss of Life in the United States Associated with Recent Atlantic Tropical Cyclones. *Bulletin of the American Meteorological Society* 81: 2065–2073.

Reinike, I. n.d. *Miss Camille 1969: The Devastating Female*. Long Beach, MS: Fennel's Coast Litho Printing Co.

Romans, B. 1775 (1999). *A Concise Natural History of East and West Florida*. Gretna, LA: Pelican Publishing Co.

Roth, D. 2010. Louisiana Hurricane History. http://www.srh.weather.gov/images/lch/tropical/lahurricanehistory.pdf, accessed, November 4, 2010.

Chapter 7

Duke, C. 2004. A Spatial and Temporal Analysis of Winter Weather Events in the Southeast U.S. with Correlations to ENSO and Other Teleconnections. M.S. Thesis, Mississippi State University.

Wax, C. L. 2008. Late freeze impacts in Mississippi. In *The Easter Freeze of April 2007: A Climatological Perspective and Assessment of Impacts and Services*. NOAA/USDA Technical Report 2008-01. http://www1.ncdc.noaa.gov/pub/data/techrpts/tr200801/tech-report-200801.pdf

Chapter 8

Elsner, J. B. 2008. Hurricanes and Climate Change. *Bulletin of the American Meteorological Society* 89: 677–679.

Grice, G. K. 2006. History of Weather Observations, Natchez, Mississippi, 1849–1955. Station History Report to the Climate Database Modernization Program of NOAA's National Climatic Data Center, Midwestern Regional Climate Center. http://mrcc .sws.uiuc.edu/FORTS/histories1.jsp#citation, accessed February 5, 2010.

Gutzler, D., R. Rosen, D. Salstein, and J. Peixoto. 1988. Patterns of Interannual Variability in the Northern Hemisphere Wintertime 850 mb Temperature Field. *Journal of Climate* 1: 949–964.

Hathorn, J. H. 2008. Paleoecology and Paleotempestology of the Pascagoula Marsh, Mississippi. Master's Thesis, Louisiana State University. http://etd.lsu.edu/docs/ available/etd-11122008-203810, accessed November 12, 2010.

Hurrell, J. 1996. Influence of Variations in Extratropical Wintertime Teleconnections on Northern Hemisphere Temperature. *Geophysical Research Letters* 23: 665–668.

Intergovernmental Panel on Climate Change (IPCC). 2007. *Contribution of Working Group I to the Fourth Assessment Report of the Intergovernmental Panel on Climate Change*, S. Solomon, D. Qin, M. Manning, Z. Chen, M. Marquis, K. B. Averyt, M. Tignor, and H. L. Miller (eds.) Cambridge, U.K.: Cambridge University Press. http:// www.ipcc.ch/publications_and_data/ar4/wg1/en/contents.html, accessed December 13, 2010.

Lisiecki, L., and M. Raymo. 2005. A Pliocene-Pleistocene Stack of 57 Globally Distributed Benthic ^{18}O Records. *Paleoceanography* 20: 1003.

Liu, K. 2007. Uncovering Prehistoric Hurricane Activity Examination of the Geological Record Reveals Some Surprising Long-Term Trends. *American Scientist* 95: 126–133.

Moran, J. M. 2006. *Weather Studies: An Introduction to the Atmosphere.* 3rd ed. Boston, MA: American Meteorological Society.

National Climate Data Center (NCDC). Global Warming: Frequently Asked Questions. http://www.ncdc.noaa.gov/oa/climate/globalwarming.html, accessed December 13, 2010.

Paulson, O. L., Jr. 1974. Physical Features: Geology. In *Atlas of Mississippi*, R. D. Cross and R. W. Wales (eds.). Jackson: University Press of Mississippi, pp. 8–11.

U.S. Global Change Research Program. 2009. Regional Climate Impacts: Southeast. In *Global Climate Change Impacts in the United States*, T. R. Karl, J. M. Melillo, and T. C. Peterson (eds.). New York: Cambridge University Press, pp. 111–116. http://www .globalchange.gov/images/cir/pdf/southeast.pdf, accessed December 13, 2010.

Chapter 9

Cathey, H. M., and R. Jordan. 2001. *USDA Plant Hardiness Zone Map.* USDA Miscellaneous Publication No. 1475. Washington, D.C.: U.S. National Arboretum Agricultural Research Service, U.S. Department of Agriculture. http://www.usna.usda.gov/ Hardzone/hrdzon2.html, accessed October 8, 2010.

Foster, G. 2004. *American Houses: A Field Guide to the Architecture of the Home.* New York: Houghton Mifflin.

Gulf Coast Business Council Research Foundation. 2008. Mississippi Gulf Coast 3.0: Three Years after Katrina. http://www.mississippirenewal.com/documents/ThreeYe arReport.pdf, accessed October 8, 2010.

Johnson, J. R.., H. Bloodworth, and D. Summers. 2002. *Agricultural Land and Water Use in Mississippi, 1992–1998*. Bulletin 1118. http://msucares.com/pubs/bulletins/b1118 .pdf, accessed August 9, 2011.

Kinton, T., and B. Dance. 2002. *Fishing Mississippi*. Jackson: University Press of Mississippi.

Lewis, D. C., and S. C. Kitchens. 2006. A Sustainable House for the Southeastern United States. http://fwrc.msstate.edu/housing/images/sustainable.pdf, accessed May 13, 2011.

Ritchie, G. L., C. W. Bednarz, P. H. Jost, and S. M. Brown. 2007. Cotton Growth and Development. http://ugakr.libs.uga.edu/handle/10724/12192, accessed August 10, 2011.

Sanders, T. n.d. Architecture in Mississippi: From Prehistoric to 1900. http://mshistory .k12.ms.us/articles/327/architecture-in-mississippi-from-prehistoric-to-1900, accessed October 8, 2010.

Von Herrman, D., R. Ingram, and W. C. Smith. 2000. Gaming in the Mississippi Economy: A Marketing, Tourism, and Economic Perspective. http://www.usm.edu/ dewd/pdf/Gamingstudy.pdf, accessed October 8, 2010.

Chapter 10

Fiebrich, C. A. 2009. History of Surface Weather Observations in the United States. *Earth-Science Reviews* 93: 77–84.

Fulton, R. B. 1892. Report of the Chief Signal Officer. In *Annual Report of the Secretary of War, U.S. War Department*. Washington, D.C.: Government Printing Office, pp. 273–296.

Grice, G. K. 2006a. History of Weather Observations, Columbus, Mississippi, 1849–1955. Station History Report to the Climate Database Modernization Program of NOAA's National Climatic Data Center, Midwestern Regional Climate Center. http://mrcc .sws.uiuc.edu/FORTS/histories1.jsp#citation, accessed February 5, 2010.

Grice, G. K. 2006b. History of Weather Observations, Jackson, Mississippi, 1849–1955. Station History Report to the Climate Database Modernization Program of NOAA's National Climatic Data Center, Midwestern Regional Climate Center. http://mrcc .sws.uiuc.edu/FORTS/histories1.jsp#citation, accessed February 5, 2010.

Grice, G. K. 2006c. History of Weather Observations, Natchez, Mississippi, 1849–1955. Station History Report to the Climate Database Modernization Program of NOAA's National Climatic Data Center, Midwestern Regional Climate Center. http://mrcc .sws.uiuc.edu/FORTS/histories1.jsp#citation, accessed February 5, 2010.

Grice, G. K. 2006d. History of Weather Observations, Vicksburg, Mississippi, 1849–1955. Station History Report to the Climate Database Modernization Program of NOAA's National Climatic Data Center, Midwestern Regional Climate Center. http://mrcc .sws.uiuc.edu/FORTS/histories1.jsp#citation, accessed February 5, 2010.

Henson, R. 1990. *Television Weathercasting: A History*. Jefferson, NC: McFarland and Co.

Moran, J. M. 2006. *Weather Studies: Introduction to Atmospherics Science*. 3rd ed. Boston, MA: American Meteorological Society.

National Weather Service Public Affairs Office. 2010. History of the National Weather Service. http://www.weather.gov/pa/history/index.php, accessed December 13, 2010.

Truesdell, R. T., J. Cooper, E. Freeman, and D. L. O'Connell. 2007. The Forms Tell a Tale: Unique Weather Observing Practices of the 19th Century. 87th Annual Meeting of the American Meteorological Society, extended abstract. http://ams.confex.com/ams/87ANNUAL/techprogram/paper_116357.htm, accessed December 13, 2010.

INDEX